MILITARY AVIATION IN NORTHERN IRELAND

An Illustrated History – 1913 to the Present Day

GUY WARNER & ERNIE CROMIE

COLOURPOINT BOOKS

Published 2012 by Colourpoint Books
An imprint of Colourpoint Creative Ltd
Colourpoint House, Jubilee Business Park
Jubilee Road, Newtownards, BT23 4YH
Tel: 028 9182 6339
Fax: 028 9182 1900
E-mail: info@colourpoint.co.uk
Web: www.colourpoint.co.uk

First Edition
First Impression

A catalogue record for this book is available from the British Library.

Designed by April Sky Design, Newtownards
Tel: 028 9182 7195 • Web: www.aprilsky.co.uk

Printed by GPS Colour Graphics Ltd, Belfast

ISBN 978-1-78073-038-7

Front cover: Military aviation at Aldergrove in the 21st century – a Gazelle of 665 Squadron AAC, with an Islander of 651 Squadron AAC in the background. *(Ulster Aviation Society)*
Small images (l-r): A No 105 Squadron RE8 overflies Omagh. *(JM Bruce/GS Leslie Collection)*
A formation of No 245 Squadron Hurricanes over Co Antrim in October 1940. *(Ernie Cromie Collection)*
The Buccaneer arrives at Langford Lodge with Gartree church in the background. *(Ulster Aviation Society)*

Rear cover: A flypast at Aldergrove in March 2007 to mark the disbandment of 655 Squadron AAC. *(Ulster Aviation Society)*
US personnel based at Langford Lodge during the war produced some very attractive samples of graffiti, now sadly destroyed. *(Ernie Cromie Collection)*
Bernadine was the mascot of 719 NAS which was based at Eglinton in the 1950s. *(Captain David James)*

CONTENTS

ABOUT THE AUTHORS

Guy Warner is a retired schoolteacher and former civil servant, who grew up in Newtownabbey, attending Abbots Cross Primary School and Belfast High School before going to Leicester University and later Stranmillis College. He now lives in Greenisland, Co Antrim with his wife Lynda. He is the author of some 20 books and booklets on aviation and has written a large number of articles for magazines in the UK, Ireland and the USA. He also reviews books for several publications, gives talks to local history societies, etc and has appeared on TV and radio programmes, discussing aspects of aviation history. He is the Vice-Chairman of the Ulster Aviation Society – for more information about the Society please see: www.ulsteraviationsociety.org

Ernie Cromie's almost lifelong interest in all things aeronautical was initially inspired by the sight and sound of a Royal Air Force Halifax bomber from the Meteorological Squadron at Aldergrove circling over farm livestock trapped by deep snow drifts around his home, in the foothills of Slieve Croob in Co Down, during the unforgettably bad winter of 1947. Twelve years later, when he was a member of the Army Cadet Force, his first-ever flight was in a RAF Anson aircraft at Jurby, Isle of Man. Unable to realise his boyhood ambition to become a pilot in the RAF because of defective eyesight, he had a fulfilling career as a town planner in the Northern Ireland Civil Service but never lost his enthusiasm for all things aviation. He has been a member of the Ulster Aviation Society since 1979, serving as its Chairman for 30 years until 2012 when he relinquished all executive responsibility in order to devote more time to researching the history of aviation in Northern Ireland, with particular reference to the United States Army Air Force and Naval Air Service presence there during the Second World War. He has written numerous articles for a wide range of aviation journals and other publications but this is his first book.

If this book encourages you to take a deeper interest in aviation in Northern Ireland, past and present, you may wish to join the **Ulster Aviation Society**. You can explore the information, history, events and aircraft owned by the society at:

www.ulsteraviationsociety.org

FOREWORD

NORTHERN IRELAND'S ASSOCIATIONS with the Royal Air Force in particular, and military aviation in general, go back a long way, a fact to which the following pages bear spectacular and poignant testimony. As an Ulsterman, it saddens me that there is no longer an RAF station in the country and I can only hope that this book may give the powers-that-be pause for thought.

My very first flight took place at RAF Aldergrove when I was at Campbell College in Belfast. The 502 (Ulster) Squadron Commanding Officer gave an illustrated lecture to the school and that really instilled an ambition to become a pilot, so I went up in a Vickers Virginia night bomber in 1933 and never looked back!

I was by no means the only Old Campbellian to serve in the Royal Air Force, which I joined in 1936. My older brother, Larmor, who was killed on 2 January 1940 when flying a Wellington of No 149 Squadron, Bomber Command, was too and during the Second World War, when I was a pilot in No 120 Squadron Coastal Command at Nutts Corner, I served alongside Eric Esler, Brian Bannister and Jack Harrison, to name but four. No 120 Squadron was of course the first in the RAF to be equipped with the B24 Liberator, which proved to be the most effective type of aircraft used by Coastal Command in our war against the U-boats, a point which I am pleased to note has not escaped the authors of this book.

The Liberator Mk I was very different from any other aircraft I had flown, including the Boeing B17C, the first of which I ferried from the west coast of the USA to Scotland in 1941, while on my first rest period. The latter flew like a big Anson! The Lib on the other hand was a very different proposition. It was heavy on the controls, especially at the overload gross weights we had to use on take-off for a 16/17 hour sortie carrying eight depth charges, or when taking high speed evasive action during Ju 88 attacks. I could never understand why four 20 mm cannons were fitted in the nose under the flight deck, as we were never allocated armour-piercing ammunition which was necessary to penetrate a U-boat's casing when it was surfaced.

The above applied also to the Sperry gyro stabilised drift sight which was quite useless for our low level operation and was quickly removed. There were no gun turrets (these came on the Liberator Mk II, III, V and VI models) on the Lib I, which was defended only by three pairs of Browning .300 machine guns, the ammo being in canisters and not belt fed. However, I am not running the aircraft down at all – it proved to be the best long range anti-U-boat machine ever built and performed its duties superbly.

The history of aviation in Ulster is extensive yet insufficiently appreciated and this account of the military aspects of it is long overdue. I congratulate the authors and publishers on their achievement and thoroughly commend it.

Squadron Leader Terence Malcolm Bulloch
DSO & Bar, DFC & Bar, RAF (Retd)

October 2012

Flight Lieutenant Terry Bulloch and his father Samuel, outside the family home in Belfast *(Ernie Cromie Collection)*

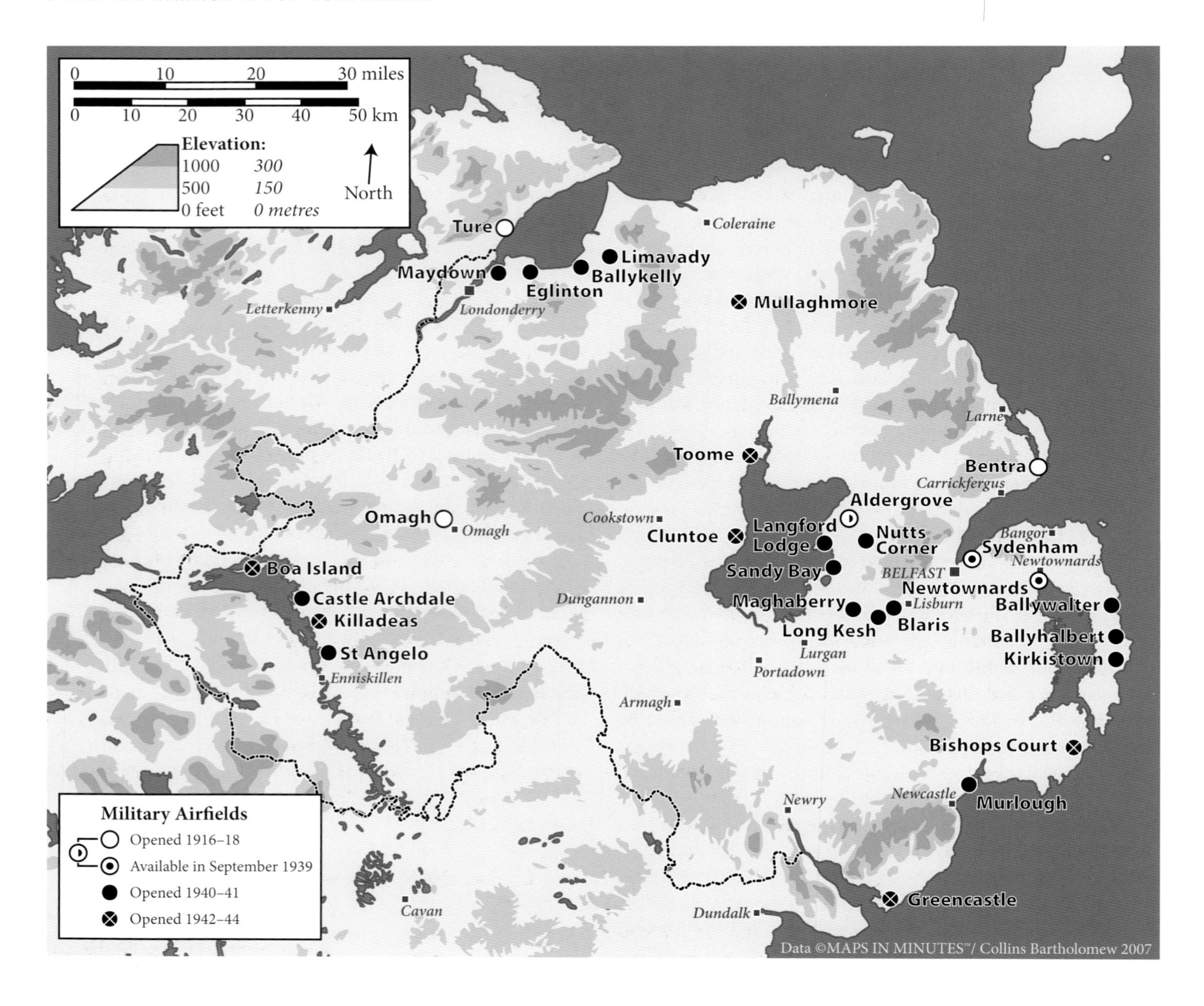
0 10 20 30 miles
0 10 20 30 40 50 km
Elevation:
1000 300
500 150
0 feet 0 metres
North
Ture
Maydown
Eglinton
Ballykelly
Limavady
Coleraine
Mullaghmore
Letterkenny
Londonderry
Ballymena
Larne
Toome
Bentra
Carrickfergus
Aldergrove
Omagh
Omagh
Cookstown
Cluntoe
Langford Lodge
Nutts Corner
Bangor
Sydenham
Newtownards
Sandy Bay
BELFAST
Newtownards
Boa Island
Castle Archdale
Killadeas
Dungannon
Maghaberry
Lisburn
Ballywalter
Blaris
Long Kesh
Ballyhalbert
Kirkistown
St Angelo
Lurgan
Portadown
Enniskillen
Armagh
Bishops Court
Military Airfields
Opened 1916–18
Available in September 1939
Opened 1940–41
Opened 1942–44
Newry
Newcastle
Murlough
Cavan
Dundalk
Greencastle
Data ©MAPS IN MINUTES™/ Collins Bartholomew 2007

INTRODUCTION

WE WERE DELIGHTED to be asked by Malcolm Johnston of Colourpoint to produce a short series of books on the history of aviation in Northern Ireland. This country has an aviation heritage which bears very favourable comparison with other regions of the British Isles, yet it is evident from the comments we receive from visitors to the Ulster Aviation Collection at the former Second World War airfield at Long Kesh near Lisburn that large numbers of the general public are less than completely well informed of the scope and importance of numerous aspects of it.

This volume is the first of three and concerns the story of military aviation in the Province over the last 100 years. It will be followed by works on civil aviation and then aircraft and aerospace manufacturing. Given the wealth of photographs available to us through our personal collections, those of friends and fellow members of the Ulster Aviation Society, and from the Public Record Office of Northern Ireland, as well as from the archives of the Ulster Aviation Collection, we have decided that the most accessible format in which to tell the story is in pictorial form, accompanied by extended explanatory captions. And what a story it is! The photographs are arranged in broadly chronological order and will take the reader on a journey in time and also around the country in peace, war and civil strife. It begins in 1913 with a truly historic event and over the next century depicts the remarkable progress that military aviation has made in our skies since the frail biplanes of the Royal Flying Corps first landed at Newcastle, Co Down. We progress through the inter-war years to the magnificent flying-boats which played such an important role in winning the Battle of the Atlantic, then to the jet age and finally to the multiplicity of helicopters which have been prevalent over the last 30 years. Aircraft from the Royal Flying Corps (RFC), the Royal Naval Air Service (RNAS), the Royal Air Force (RAF), the Fleet Air Arm (FAA) and the Army Air Corps (AAC) are all featured, as indeed are aircraft from our neighbours, the Irish Air Corps and the wartime United States Army Air Force (USAAF). We have also depicted locations and people as the story of aviation is not just about flying machines.

We hope that you will enjoy looking at and reading this chronicle as much as we have enjoyed compiling it and look forward to working on the next two volumes.

We would also like to express our sincere thanks to fellow members of the Ulster Aviation Society for their advice and help with specific aspects, especially the Chairman, Ray Burrows, Wing Commander Eddie Cadden RAF Voluntary Reserve Training (VRT) Retd and Brigadier General Paul Fry of the Irish Air Corps.

It would be remiss of us not to mention our gratitude for the support we have received over many years from our tolerant or perhaps, long-suffering, wives, Lucy and Lynda.

Ernie Cromie, Hillsborough
Guy Warner, Carrickfergus

October 2012

A photograph from the *Belfast Evening Telegraph* of 2 September 1912. *(Belfast Central Newspaper Library)*

No 2 Squadron at Limerick in September 1913. *(Guy Warner Collection)*

On Monday 1 September 1913, in what was, effectively, the first overseas deployment of the Royal Flying Corps (RFC), five BE2a aircraft, serial numbers 217, 218, 272, 225 and 273, flown by Captains JWH Becke, CAH Longcroft and ACH McClean, Lieutenants FF Waldron and L Dawes respectively, and a single Maurice Farman Longhorn, No 207, piloted by Captain GWP Dawes, all of No 2 Squadron based at Montrose, flew from Scotland to Limerick, to take part in large-scale Irish Command manoeuvres. About 1.50 pm, to refuel and carry out some essential maintenance, Captain Dawes landed his Longhorn on the beach near the Slieve Donard Hotel, Newcastle and took off about an hour later for Limerick where he re-joined his colleagues that evening. Near the end of the month four BE2s flown by Longcroft, McClean, L Dawes and Lieutenant RB Martyn broke their return journey at Newcastle to re-attach floatation gear in advance of the sea crossing. The first three named landed near Bryansford, while Martyn alighted at Dundrum. It was noted that the airmen were "royally entertained" to lunch in the clubhouse of the Royal Co Down Golf Club, while Longcroft also performed a flying display over the town.

The non-rigid airship SS23 lands at Bentra in 1917. *(D&N Calwell Collection)*

Ground crew in front of the hangar at Bentra. *(D&N Calwell Collection)*

Between 1914 and 1918, during the First World War, 23 airfields and airship bases were established in the whole of Ireland for the RFC, RNAS, US Naval Air Service and RAF, which was formed on 1 April 1918 by the amalgamation of the RFC and RNAS. The first dedicated aviation facility in Ulster was an airship mooring-out station at Bentra, near Whitehead in Co Antrim, which was provided by the RNAS, Luce Bay at Stranraer being its parent operating base for coastal patrol airships of the SS and later SSZ classes. A regular duty of the airships was to escort the Larne–Stranraer ferry, *Princess Maud*. In case of unfavourable weather, the facilities provided at Bentra included a 150 feet long portable airship shed, which was used extensively during the last two years of the war.

A Handley Page V/1500 at Aldergrove. *(Ernie Cromie Collection)*

During May/June 1917, the farmland which became part of Aldergrove airfield was surveyed by Major Sholto Douglas RFC and declared to be a suitable site for the establishment of a flying training school for the Corps. In the event it became instead No 16 Aircraft Acceptance Park when Harland and Wolff gained a contract to manufacture Handley Page V/1500 four-engine heavy bombers. Aldergrove was used to test fly the aircraft, the first of which, E4304, flew from there on 20 December 1918, piloted by the Handley Page test pilot, American-born Clifford Prodger. Whether the aircraft were fully constructed or simply re-assembled at Aldergrove after being brought by road or railway from Belfast is not known for certain. The hangar/workshop in which they were assembled at Aldergrove was nearly 600 feet long and 150 feet wide, and was directly linked to the Lisburn–Antrim section of the Great Northern Railway by a branch rail track one and a half miles in length. The hangar was seriously damaged by a storm during the 1920s and was only partially re-built, but still exists and is now a listed building of architectural and historic interest.

A No 105 Squadron RE8 overflies Omagh. *(JM Bruce/GS Leslie Collection)*

Strathroy Aerodrome, near Omagh. Note the canvas hangars. *(Ernie Cromie Collection)*

Meanwhile, in May 1918, No 105 Squadron RAF, which was equipped with two-seat, RE8 aircraft, was sent to Strathroy airfield, near Omagh. Its duties were to support the Army on the ground on reconnaissance and communications tasks. The aim, in the words of the Lord Lieutenant of Ireland, Field Marshal Viscount French, was to "put the fear of God into these playful Sinn Feinners." It remained there until early 1919. Accommodation for aircraft and personnel based at Strathroy was provided in the form of tents and nothing remains of the facility. The RE8 was designed at the Royal Aircraft Factory, Farnborough and replaced the obsolescent BE2 variants which had given yeoman service since 1914. It was responsible for by far the greater part of artillery observation work on the Western Front in France from late 1916 onwards. More than 3000 RE8s were constructed.

Handley Page 0/400, J2259 at Baldonnel in 1920 – now the HQ of the Irish Air Corps. *(JM Bruce/GS Leslie Collection)*

Many long distance flights were attempted by the RAF between the wars but first, aircraft reliability and the crew's navigational skills needed to be tested. Major Keith Park (of subsequent Battle of Britain fame) carried out a remarkable endurance flight in a Handley Page 0/400 in April 1919. Flying F3750 *Last Days* of No 1 Air Navigation and Bombing School, he circumnavigated the British Isles. The flight began from Andover in Hampshire. The first leg followed the east coast route to Edinburgh, the next day they rounded the north of Scotland and headed down the west coast for the sea crossing to Aldergrove. When they arrived, the airfield was fog-bound, it was 7 pm and their fuel was running low. Park completed a masterly feat of airmanship by bringing off a crosswind landing at Queen's Island on a Harland and Wolff wharf which measured only 400 yards by 50 yards. The rest of the crew went to Aldergrove by road. Park took off the next morning, picked them up and continued the circuit via Dublin, Pembroke, Bodmin, Plymouth and Bournemouth, returning to Andover after four days in which the aircraft had covered 1600 miles in 30 hours in the air.

Aldergrove main gate 1920.
(Ernie Cromie Collection)

Bristol Fighter, F4796, of No 106 Squadron, at Collinstown (now Dublin Airport) in 1919.
(JM Bruce/GS Leslie Collection)

Whereas the reason for the existence of No 16 Aircraft Acceptance Park had ceased by 1920, Aldergrove was retained as an airfield and a detached Flight of Bristol Fighters from No 4 Squadron RAF was stationed there from 1920 to 1922. The Flight was part of several RAF units which were based in Ireland between 1919 and 1923, including No 2 Squadron with similarly equipped Flights at Oranmore, Fermoy and Castlebar and No 100 Squadron, initially equipped with Handley Page O/400s and DH9s, based at Baldonnel, with detached Flights at Castlebar and subsequently Oranmore with Bristol Fighters and DH9s. Communications with No 11 (Irish) Group Headquarters in Dublin were maintained by No 100 Squadron until January 1922 when the squadron was transferred to England. With the signing of the Anglo-Irish Treaty in January 1922, the RAF began the process of withdrawal from what would become the Irish Free State in December 1922, but to protect the evacuation an Irish Flight of four Bristol Fighters was temporarily retained at Baldonnel. On 31 May 1922, No 2 Squadron, then based in England, was moved to Aldergrove with 12 Bristol Fighters, their principal task being defined as "establishment of an aerial mail between Belfast and Dublin and preparations to co-operate with various brigades and patrols along the Ulster border." The mail service between Dublin (Collinstown) and Aldergrove began on 7 June and was flown at a rate of up to two aircraft a day until 23 October. Aircraft from No 2 Squadron and the Irish Flight were used. The aeroplanes' machine guns were either removed or rendered inoperable for flights over Irish territory. To cover the final stages of withdrawal of British forces, a detachment of No 2 Squadron aircraft was meanwhile sent from Aldergrove to Collinstown in June, fully armed with bombs and ready to act if required. The run-down of British forces in Ireland resulted in changes to the Command structure, not least in the RAF. By September 1922, No 11 (Irish) Group had become No 12 (Irish) Wing, the Headquarters of which moved to Northern Ireland, initially to Castle Upton, Templepatrick in September, from whence it moved on 30 November to Aldergrove, where it disbanded in February 1923 and the airfield was placed in Care and Maintenance.

Vickers Vimys of No 502 Squadron at Aldergrove in April 1927. *(Ernie Cromie Collection)*

Vickers Vimy of No 502 Squadron. Note the red hand badge. *(Ernie Cromie Collection)*

By 1920, the RAF had been reduced to one tenth of its peak strength around the time of its formation in 1918. Concern about its ability to fulfil its commitments eventually gave rise to the Air Force and Air Force Reserve Act of 1924. This resulted, throughout the UK during the 1920s and 1930s, in the formation of Special Reserve and Auxiliary Air Force flying squadrons, five in the former category and 15 in the latter, respectively. A squadron of the Special Reserve was commanded by a regular RAF officer and consisted of a nucleus, or cadre, of regulars. About half of the squadron consisted of part-time volunteers drawn from its local catchment area. Auxiliary squadrons were also 'territorial' in ethos but were almost wholly manned by part-time volunteers. The first of these 20 squadrons to be formed, in May 1925, was No 502 (Ulster) Squadron, at Aldergrove. Originally it was a Special Reserve unit, with a day-bombing role and was part of the Air Defence of Great Britain. The new squadron first occupied the hangars and buildings formerly used by No 16 AAP and a recruiting office in Belfast was opened in July. The first aircraft on strength were Avro 504 twin-seat biplane trainers and the much larger Vickers Vimy twin-engine bombers.

Supermarine Southampton Mk 1, N9899, moored off Harland and Wolff's shipyard in 1925. The hull is preserved on display at the RAF Museum, Hendon. *(Ernie Cromie Collection)*

In September 1925 Flight Lieutenant Shoppe and Flight Lieutenant JH Bentham flew the Supermarine Southampton N9899 of No 480 Coastal Reconnaissance Flight to Northern Ireland from Calshot by way of Portland, Cattewater and Pembroke Dock towards Belfast. However, engine failure 15 miles south-east of Wicklow Head forced them to alight on the water. They were rescued and towed to Belfast by the light cruiser HMS *Calliope*. Repairs were effected on one of the slipways at Harland and Wolff's shipyard and a trial flight was made from Belfast Lough on 21 September. The *Belfast Evening Telegraph* reported that a 'flotilla' of similar flying-boats alighted on the Lough that same day. All these flights were part of a series of round-Britain training and flag-showing 'cruises'. The officers were brought ashore at Carrickfergus in the motor launch *Zaida* and the next day they flew on by way of a touch-down on Lough Neagh, not far from the RAF airfield at Aldergrove.

Vickers Vimy over Co Down. *(J Hewitt Collection)*

Another example of 'flag-showing' took place on 25 March 1926. With the aim of encouraging recruiting as well as demonstrating No 502 Squadron's growing competence, a flight by four Vimy aircraft was made around Ulster, led by the Commanding Officer, Squadron Leader RD Oxland, carrying members of the press. After circling around the shores of Lough Neagh for about 15 minutes, the machines assembled into formation and headed straight for Belfast, in 10 minutes time Bangor was reached. Heading south down the Ards Peninsula and passing over Newtownards and Downpatrick, they proceeded along the coast to Newcastle. They then turned sharply inland and steered for Armagh. The flight continued over Co Tyrone, passing over Dungannon, Sixmilecross and Omagh. Crossing into Co Fermanagh, they overflew Enniskillen before heading towards Newtownstewart and following the River Foyle to a point near Derry, turning eastwards they passed over Limavady and were soon near Coleraine. They followed the course of the River Bann to Lough Neagh and continued over Randalstown and Antrim. Following a gradual descent they reached terra firma at Aldergrove again after three and a half hours in the air. Two months later, the squadron was authorised to use the 'Red Hand of Ulster' as the symbol on its badge and aircraft.

Handley Page Hyderabad and Avro 504Ns. *(Ernie Cromie Collection)*

Handley Page Hyderabads on Tyrella beach Co Down. *(J Hewitt Collection)*

The robust and reliable Vimy served with No 502 Squadron until 1928, when it was replaced by the Handley Page Hyderabad, the last heavy bomber of wooden construction to serve with the RAF. Annual summer camps at an airfield in Great Britain and cross-country flights which were two aspects of the squadron's training were not always accident-free. On 27 August 1929 for instance, three of its Hyderabads were returning from Manston in Kent, via a refuelling stop at Sealand in Cheshire when they encountered conditions of bad visibility after crossing the Irish Sea. One managed to reach Aldergrove safely; the other two decided to land at Groomsport in a field used occasionally for pleasure-flying operations by the North British Aviation Company. One landed safely but the other, J8814, piloted by the Officer Commanding (OC) 'A' Flight, Squadron Leader CL King with six personnel on board, ran into some trees on landing and suffered major damage, beyond repair. Remarkably, all on board survived, without suffering serious injury. They included Flying Officer AM Stevens who, following a long career in the RAF, retired to live in Bangor, Co Down.

Vickers Virginia at Aldergrove. *(J Hewitt Collection)*

In turn, the Hyderabad was replaced in 1931 by the Vickers Virginia. On 16 November 1932, a formation of Vickers Virginias of No 502 Squadron took off from Aldergrove to provide the Prince of Wales with an aerial escort as he arrived in Belfast for the opening of the new parliament buildings at Stormont. Before they were retired towards the end of 1935, the Virginias took part in a further Royal ceremony, when the Duke of Gloucester, who was also Earl of Ulster, paid a three-day visit to Northern Ireland in May that year, representing the King and Queen as part of the Silver Jubilee celebrations. Three aircraft escorted HMS *Achilles*, as the cruiser brought the Duke up Belfast Lough. Despite its appearance, the Virginia achieved an enviable reputation as a very steady bombing aircraft which pioneered both the automatic pilot and the use of a gunner's position in the extreme tail. Separated from the remainder of the flight crew by a long fuselage through which he had no means of access, it is said that when the rear gunner of a Virginia wished to attract the pilot's attention he reached out and waggled the elevator!

Some of General Balbo's flying-boats in Rosses Bay, Lough Foyle. *(Belfast Telegraph)*

Lord Londonderry with General Balbo. *(Ernie Cromie Collection)*

On 2 July 1933, 24 Savoia Marchetti S55x flying-boats of the Italian Air Force, under the command of General Italo Balbo, en-route to Chicago and the first crossing in formation of the North Atlantic, alighted on Lough Foyle. They were escorted on the final leg of their transit to Lough Foyle by four Southamptons of No 201 (Flying-Boat) Squadron and, on the following day, the Secretary of State for Air, Lord Londonderry arrived at Aldergrove from London in a Hawker Hart of No 24 Squadron RAF. From there he was conveyed to Lough Neagh, where he boarded a Southampton in which he flew to Lough Foyle to greet General Balbo. Balbo's air armada was moored on Lough Foyle for three days, during which time he was the guest of Lord Londonderry at Mount Stewart, the family seat on the eastern shore of Stangford Lough, not far from Newtownards, where the photo was taken on the occasion.

A No 502 Squadron Westland Wallace preparing for flight. *(J Hewitt Collection)*

In October 1935, the role of No 502 Squadron was changed from a twin engine bomber squadron to a 'Cadre Bomber Squadron – single engine', which resulted in the Virginias being replaced by the general purpose Westland Wallace. Two early versions of the type, specially modified and piloted by the Commanding Officer and Second-in-Command of No 602 (City of Glasgow) Squadron, with their observers, had flown over Mount Everest in April 1933, the first aircraft to do so. Powered by a 550 hp Bristol Pegasus radial engine, the standard military version could achieve a maximum height of 24,000 feet and carry a bomb load of between 580 and 1000 pounds on underwing mountings.

A Gloster Gauntlet of the Met Flight in 1937. *(Ernie Cromie Collection)*

1936 saw several significant developments at Aldergrove. On 28 January it officially became a Royal Air Force Station, under the command of Group Captain JC Russell Distinguished Service Order (DSO). Another was the establishment of a Meteorological Flight. This was the second such unit to be formed in the RAF, with the first being at Duxford in Cambridgeshire. The unit was equipped with Bristol Bulldogs, which had first been introduced into service in 1929. Considering that the job was to go aloft in all weathers to 20,000 feet, the pilots were a hardy and special breed. The first officer in command of the Met Flight was of that kind, Flight Lieutenant Victor Beamish of Coleraine. The aim of the Met Flight was not only to find out what the atmospheric conditions were but also to help with the study of weather systems which built up in the Atlantic and so to develop a greater understanding of the patterns of depression and anti-cyclone which dominate our lives and conversation. The Bulldogs were replaced in July 1937 with the more powerful Gloster Gauntlet, a type which has the distinction of being the last of the RAF's long line of open-cockpit, biplane fighters. It should be noted that a civilian meteorological unit under the direction of JD Ashton and his staff of three had been formed at Aldergrove in July 1926.

The Killultagh Foxhounds at Aldergrove with two Heyfords on 22 October 1936. *(PRONI D2334/6/6/7)*

A Blackburn Roc makes a simulated attack on a Fairey Swordfish south of Sandy Bay. *(Ernie Cromie Collection)*

The diver is Joe Magee who worked for the McGarry family. *(Ernie Cromie Collection)*

In March 1936, as part of its expansion programme due to the increasing likelihood of another European war, the RAF set up an Armament Training Camp at Aldergrove. In the early 1930s, to meet the needs of No 502 Squadron, fairly rudimentary floating target structures had been created on the eastern side of Lough Neagh by Henry McGarry of Ardmore Boatyard. With their associated land-based facilities, they were developed and expanded to provide for practice and live bombing, as well as air to air and air to ground firing. The initial establishment of No 2 Armament Training Camp was Westland Wallace aircraft. Later aircraft on the strength of the unit, which became No 2 Armament Training School (2 ATS) in 1938, included Fairey Battles and Hawker Henleys. In April No 1 ATS and No 2 ATS combined to become No 3 Air Observers School, which in turn was reconstituted as No 3 Bombing and Gunnery School in November 1939. Another type to use the facilities was the Handley Page Heyford heavy bomber.

Hawker Hinds of No 502 Squadron on patrol. *(Ernie Cromie Collection)*

In 1937, No 502 Squadron was re-equipped again, this time with the Hawker Hind, which was a development of their famous and aesthetically pleasing Hart. It was however, still a biplane, with open cockpits for its crew of two, made of fabric-covered steel and aluminium and having a top speed of 186 mph. Two Avro Tutors and a Hart were also received for use by the squadron training flight. A change in status from being Special Reserve to becoming an Auxiliary Air Force Squadron occurred on 1 July 1937. The following year, Lord Londonderry was appointed Honorary Air Commodore. Tom Grant, aged 18, joined No 502 Squadron in 1938 and trained as an air gunner. He later recalled a design peculiarity of the Hind – the bomb aimer's position, which was below the pilot's seat. He had to crawl underneath, while the pilot operated a hand crank to raise the radiator which otherwise obscured the downward view. The bomb aimer had to be swift as well as accurate, as unless the radiator was lowered again speedily, the engine would overheat.

Avro Anson, N5104, No 502 Squadron flying over Antrim in 1939. *(Ernie Cromie Collection)*

With the prospect of war looming ever closer, No 502 Squadron became part of 18 Group of Coastal Command in November 1938. Reflecting its change of role to reconnaissance, it received monoplane aircraft for the first time in January 1939, Avro Ansons and, over the next eight months, 19 of them were delivered. In June 1939 it was further transferred to 15 Group, Coastal Command. The first U-boat sighting was made in the early days of the war, on 24 September, close to Rathlin Island by the OC 'B' Flight, Flight Lieutenant Philip Billing, in Anson N5104. The submarine was attacked as it submerged but no hits were observed. This particular Anson, flown by Pilot Officer Hunter McGiffin, was pictured over Antrim prior to the outbreak of war, before being fully equipped for its offensive role. Standard armament of the Anson was one fixed forward-firing .303 inch machine gun with another in a dorsal turret while its ordnance load was four 100 pound bombs which were useless against the pressure-hulled U-boats. John Hanna, from Lisburn, who was a gunner with No 502 Squadron early in the war, recalled, "my only opportunity to open fire came when an enemy submarine was surprised on the surface. On such occasions, we would fly low over the sub, drop our bombs and the air gunners would spray the sub with machine gun fire in the hope of damaging its superstructure or hitting any gun crew should they attempt to return fire from their deck gun … on making an attack, we would immediately contact warships in the area who would move in at fast speed. With undersea detection equipment and powerful depth charges they stood a much better chance of obtaining a confirmed kill."

Ken Mackenzie (the pilot) in a Hawker Hind at Sydenham in 1939. *(Wing Commander Ken Mackenzie)*

No 24 E&RFTS personnel pose with a Hawker Hind and DH Tiger Moths at Sydenham. *(Ernie Cromie Collection)*

There were also developments at Sydenham airfield in Belfast. In January 1939 an Elementary and Reserve Flying Training School (No 24 E&RFTS) was set up for the instruction of new pilots for the Volunteer Reserve of the RAF. Its basic training aircraft was the famous DH Tiger Moth, with students progressing to more advanced Hawker Hind two-seat biplanes. The instruction and administration was run by the RAF with the servicing provided by Short and Harland. On the outbreak of war, on 3 January 1939, No 24 Elementary Flying Training School (EFTS) was formed at Sydenham out of No 24 E&RFTS and No 23 E&RFTS, Rochester. By 1 January 1940 it was equipped with 36 aircraft, Tiger Moths, Magisters and Demons and now also began training RN and RNVR pilots. In the front row of this group of pilots, fifth from left, with small moustache, is John Butler, a Belfast optometrist who later during the war, was employed by the Americans at Langford Lodge in a civilian role and emigrated to the USA in 1947. Another member of No 24 E&RFTS was Sergeant Pilot Ken Mackenzie who would attain the rank of Wing Commander, win a DFC and an AFC, fight in the Battle of Britain, and shoot down 12 enemy aircraft flying with No 501 and No 247 Squadrons.

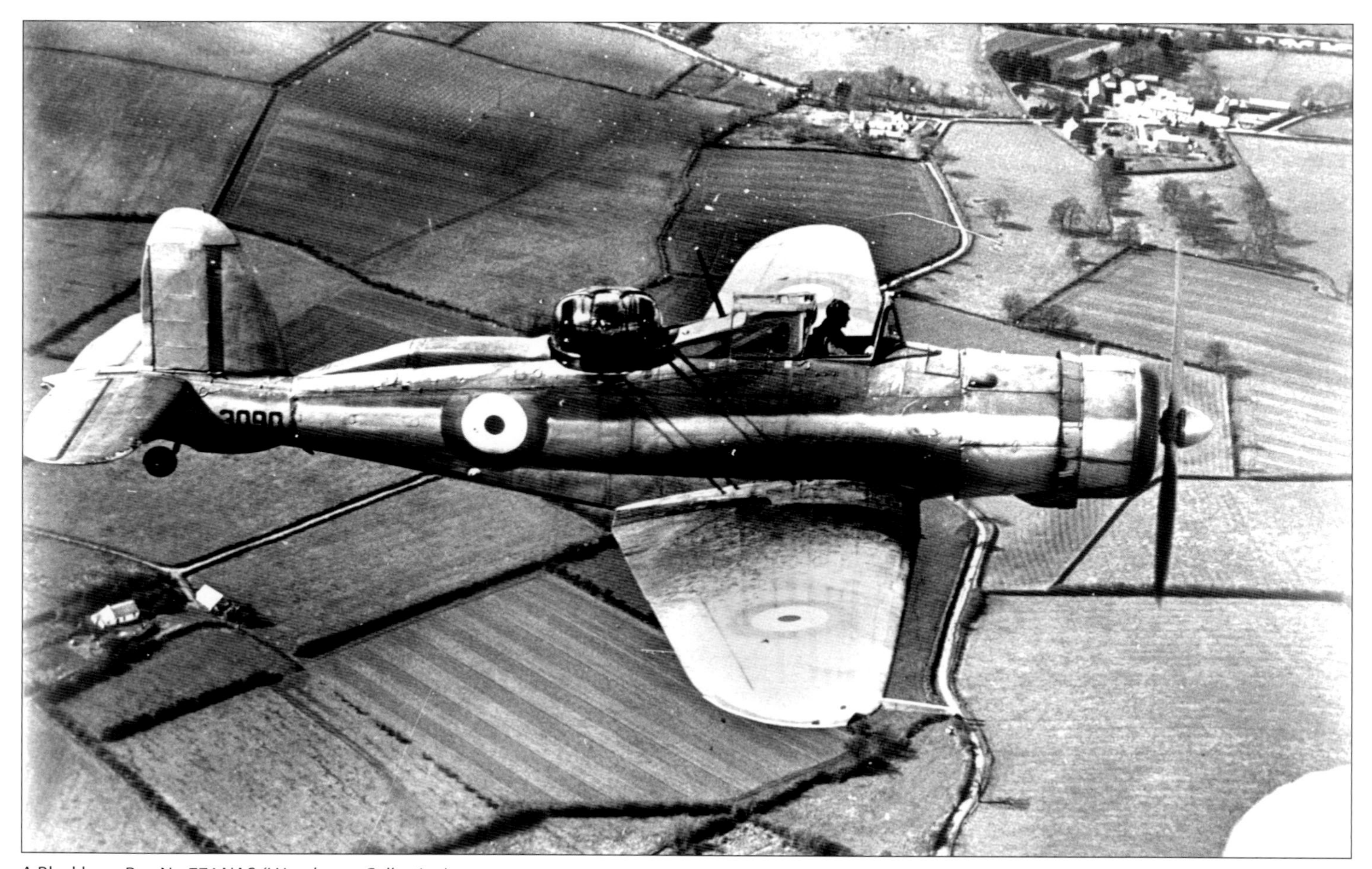

A Blackburn Roc No 774 NAS *(I Henderson Collection)*

774 Naval Air Squadron also joined No 3 Bombing and Gunnery School at Aldergrove on Christmas Day 1939 with Fairy Swordfish, Blackburn Roc, Shark and Skua aircraft to provide Telegraphist Air Gunner training. It had the distinction of being the first Fleet Air Arm squadron to be based in Northern Ireland and remained at Aldergrove until July 1940 when it left for Evanton in Scotland.

Early in 1940 the Rocs and Skuas were reorganised to form a very rudimentary fighter squadron for the defence of the province in general and Belfast in particular, as no other more suitable aircraft could be spared at that stage of the war. Thankfully they were not called into action against hostile aircraft.

Nearly all the hangars in this very sharp image taken 3 November 1941 overhead Aldergrove, were used by No 23 MU. *(Ernie Cromie Collection)*

Before an aircraft was ready for military use it had to pass from the manufacturers through a service Maintenance Unit (MU). In November 1939 No 23 MU was established at Aldergrove. The job of the MUs, which were largely civilian staffed, was to prepare aircraft for operational or second line duties, fit armament and radios or carry out other specific modifications. Aircraft also needed regular maintenance that neither squadron nor station engineering staff could provide. They were then stored to await delivery to their unit.

The first aircraft to receive the attention of 23 MU was a Bristol Bombay, one of 50 of this bomber/transport built under contract by Short & Harland. These were the first aircraft to be built at the new factory at Airport Road, Sydenham and to make their maiden flights from the adjoining airfield, the aircraft were taken across a connecting bridge. Thereafter expansion was rapid and within the first year the ten hangars occupied by the MU had received over 500 aircraft, of which more than 300 had been passed on to service duties.

Fairey Battle of No 88 Squadron at Sydenham in April 1941. *(Ernie Cromie Collection)*

In May 1940 five Hawker Demon two-seat, biplane fighters were stationed at Sydenham and flown by the instructors from No 24 EFTS for Station defence purposes on a round-the-clock availability basis. No other locally-based fighter aircraft were available at this stage for the defence of Northern Ireland. The arrival at Sydenham of two Army Co-operation squadrons of obsolescent Fairey Battle light bombers, Nos 88 and 226, which had been withdrawn from France near the end of June 1940 to re-group and rest, did not help significantly. The Battle pilots took over the Demons and responsibility for the fighter protection of the Station but they were soon relieved of this duty with the arrival at Aldergrove of No 245 Squadron, Fighter Command with Hurricanes in July, more or less at the same time as the transfer of No 24 EFTS to Luton.

CAM ship Hurricane being launched. *(via Mike Lewis)*

702 NAS Sydenham 1942. *(David Wright)*

An attempt to develop a solution to the problem posed to convoys by German long-range Focke-Wulf Condors, was the development of Fighter Catapult Ships (FCS) and Catapult Aircraft Merchant Ships (CAM) in 1941. These carried Fairey Fulmars or Hawker Sea Hurricanes which could be launched if a Condor approached a convoy. The aircraft could not land back on the ship again so either had to ditch or head for the nearest friendly territory – Northern Ireland on several occasions. The first operational launch was from HMS *Pegasus* on 11 January 1941. After chasing away the Condor, the Fulmar, flown by Petty Officer J Shaw, made a safe landing at Aldergrove. Another landed at Sydenham on 7 June, a third sadly crashed into a hillside on 7 July. On 27 August Fulmar N4072 (Sub-Lieutenant Birrell and Leading Naval Airman Sykes) was launched from HMS *Ariguani*. After engaging the enemy Birrell flew for two hours to make a landing at Tramore Strand in Co Donegal. He was able to take off again and made his way to Eglinton. In July 1941, 804 Naval Air Squadron (NAS) was attached to RAF Sydenham (which changed its name to RAF Belfast in November) until May 1942. The attachment ceased in May 1942. A flight of 702 NAS was formed with two Sea Hurricanes but never went to sea as the requirement for FCS and CAM vessels was reducing. It disbanded in July 1942. A memorial to the CAM ship pilots was unveiled by Dame Mary Peters and Mrs Norma Wright at George Best Belfast City Airport in June 2010.

Squadron Leader John Simpson, OC No 245 Squadron. *(Ernie Cromie Collection)*

The ad-hoc arrangements of the early days, with obsolete Hawker Demons being flown by instructors from Sydenham and the Bombing and Gunnery School at Aldergrove providing Blackburn Rocs and Skuas, was proven to be totally inadequate by the devastation wrought in the Belfast Blitz of 15–16 April and 5–6 May 1941. An estimated total of 900 people were killed in the first raid, while another 200 died in the second one. A total of 35 bomb craters temporarily put RAF Sydenham out of action. The provision of anti-aircraft guns and searchlights was well below what was needed to protect a major port and centre of industrial production effectively. By that time No 245 Squadron, based at Aldergrove with Hurricanes, was providing fighter cover for Northern Ireland. Although not equipped for night fighting, a few aircraft took to the air. To avoid collisions, all sorties were flown by single aircraft and two Luftwaffe bombers were shot down. In July 1941, the Squadron was transferred to Ballyhalbert, where a fighter airfield and Sector HQ had been completed. No 82 Group/Sector Operations Room was established in the Senate Chamber at Stormont in September 1941 and closed some four years later around the time of VE Day.

RAF Limavady, note Lough Foyle and hills of Donegal. *(Ernie Cromie Collection)*

One of Northern Ireland's main contributions to the war was as a base for Royal Air Force Coastal Command (RAFCC) squadrons, as they waged an unceasing and bitter struggle against the U-boat menace, actively engaged from the first day of the war to the last. Combat units operating within the RAFCC, including squadrons of the Fleet Air Arm in the closing months of the war, were based for significant periods of time at seven airfields. They were, in order of being used: Aldergrove, Limavady, Castle Archdale, Nutts Corner, Ballykelly, St Angelo and Mullaghmore. Collectively, they were the most strategically important of the country's 25 military airfields and made a vital contribution to the ultimate victory achieved in the Battle of the Atlantic. Aircraft operating from them, unassisted by surface ships of allied naval forces, were officially credited with the destruction of at least 25 U-boats – the precise figure will always be open to dispute.

A formation of Lockheed Hudsons of No 48 Squadron. *(RAF Aldergrove)*

Flight Lieutenant Riley's Beaufighter of No 252 Squadron. *(RAF Aldergrove)*

The Avro Ansons of No 502 Squadron at Aldergrove were soon succeeded in operational service from local airfields by more capable, more heavily armed and longer ranged types, such as Lockheed Hudsons, Armstrong Whitworth Whitleys, Vickers Wellingtons, Bristol Blenheims and Bristol Beaufighters, which were introduced during the first two years of hostilities. Detachments of several squadrons, including Bristol Blenheims of No 254 Squadron, as well as Lockheed Hudsons of Nos 224 and 233 Squadrons were based at Aldergrove for periods of up to several months. A new unit, No 272 Squadron formed at Aldergrove in November 1940, equipped with Blenheims. Some material evidence of success was gained by the Hudsons of 233 Squadron, with the shooting down of a Heinkel 111 and a Condor in May and July 1941, respectively, both over the Atlantic. New and more potent aircraft arrived that year in the form of the Bristol Beaufighters of Nos 143 and 252 Squadrons. A Beaufighter piloted by Flight Lieutenant Riley shot down a Condor on 16 April 1941.

Boeing Fortress II FL459 of No 220 Squadron. *(Ulster Aviation Society)*

Liberator Mk IIIs and Vs of Nos 86 and 120 Squadrons at Aldergrove. *(RAF Aldergrove)*

It was in the autumn of 1941 that the capability to patrol further out into the Atlantic, and so give greater protection to the convoys and to combat the U-boat 'Wolf-packs', was created by the introduction to operations of four-engine, Very Long Range (VLR) landplane types. Boeing B17s were used as interim types but the most effective aircraft was another American product, the formidable Consolidated B24 Liberator. For almost a year No 120 Squadron at Nutts Corner, equipped with Mark I Liberators, was the only operational VLR unit in Coastal Command. One of that squadron's pilots was Terry Bulloch from Lisburn who flew with No 120 Squadron and ended the war as a Squadron Leader with bars to both the DSO and Distinguished Flying Cross (DFC). He was the top RAFCC U-boat killer of the war, being officially credited with four destroyed and a greater number damaged. Ballykelly, which had been functioning as the home of the Coastal Command Development Unit (CCDU) since late 1941, received its own VLR aircraft in 1942, in the shape of the B17s of No 220 Squadron, replacing the CCDU. No 120 Squadron then moved from Nutts Corner to join No 220 at Ballykelly.

Very Long Range Liberator Mk 1, AM 923, which joined No 120 Squadron on 12 October 1941. *(Ernie Cromie Collection)*

Flying Officer Eric Esler and his crew with their Liberator of No 120 Squadron. *(Peter Clegg Collection)*

The most critical and sustained battle fought by the RAF during the Second World War was against the U-boats. The most successful year for the sinking of these feared adversaries was 1943, with at least 18 destroyed by RAFCC aircraft based in Northern Ireland, nearly 20% of all U-boats sunk by aircraft action exclusively in all theatres of the war during that record year, a testimony to Northern Ireland's strategic geographical location. The most effective type available to the Command was the B24 Liberator which destroyed more U-boats than all the other types operating from Northern Ireland bases. Significantly though, it was a B17 Flying Fortress of No 220 Squadron, operating from Ballykelly which was the first Northern Ireland based aircraft to have the distinction of a confirmed U-boat kill on 3 February 1943, when U-265 was sunk. Four days later U-624 was successfully attacked and dispatched by another Fortress from the same squadron. By early March, three U-boats had been sent to the bottom. No 86 Squadron, equipped with Liberators, added to the score with three more sinkings in April and May 1943, the total being made up during this critical period by a third Liberator-equipped unit, No 59 Squadron, and Short Sunderland flying-boats operating from Lough Erne.

Killadeas Flying-Boat base on Lough Erne. *(Ernie Cromie Collection)*

Training was also a vital function and RAFCC used Long Kesh, Maghaberry, Maydown, Limavady, St Angelo, Sandy Bay and Killadeas to accommodate a range of training units for varying periods. Twelve flying-boat moorings were established at Sandy Bay, in the shelter of Ram's Island, Lough Neagh, along with a number of marine craft moorings for attendant vessels and refuellers. Gas-lit navigation buoys were also laid out to guide incoming and outgoing aircraft. The most frequent visitors were Short Sunderlands and Consolidated Catalinas. Operational Training Units used two of the airfields almost exclusively, No 7 (C) OTU at Limavady with Vickers Wellingtons throughout the second half of 1942 and all of 1943, and No 5 (C) OTU at Long Kesh mainly with Bristol Beauforts and Lockheed Hudsons throughout the whole of 1943. It should be added, however, that operations were also flown from Limavady between 1940 and 1942 by Wellingtons, Whitleys and Hudsons, and again during 1944 and 1945 by Royal Navy Swordfish and Wildcats. There is more to learn of Long Kesh later.

Short Sunderlands of No 201 Squadron in the Donegal Corridor. *(Ernie Cromie Collection)*

The first flying-boat to land on Lough Erne was a Supermarine Stranraer biplane of No 240 Squadron in February 1941. Soon more modern types followed, the ill-fated Saro Lerwicks of No 209 Squadron, the slow but utterly practical Consolidated Catalina and the magnificent Short Sunderland. Two squadrons in particular, both equipped with Sunderlands, became the longest-based residents at Castle Archdale, No 201 RAF from October 1941 until April 1944, again from November 1944 until August 1945, and No 423 of the Royal Canadian Air Force (RCAF), from November 1942 until July 1945. The Éire Government was determined that Éire would remain neutral but an agreement was negotiated which allowed flying-boats based on Lower Lough Erne to fly westwards to the Atlantic along an air corridor over Ballyshannon in Éire between Lough Erne and Donegal Bay. This served to increase the effectiveness of operations by reducing transit times and increasing the range of cover that could be given to convoys in the Eastern Atlantic. Land-based aircraft operating to and from other airfields also made effective use of the Air Corridor, within which, in Northern Ireland, the Americans operated a Radio Range that was constructed just to the east of Belleek as a navigational aid for aircraft being ferried from the USA and RAFCC aircraft returning to Lough Erne from operations over the Atlantic.

Icebound Catalina on Lough Erne.
(Ernie Cromie Collection)

Catalinas and Sunderlands beside hangars at Castle Archdale.
(PRONI CAB/3/G/9/8b)

One particular example from many, and the best-known, may serve to illustrate the vital utility of the Lough Erne base. On Monday 26 May 1941, it was Catalina AH545/Z of No 209 Squadron, captained by the highly experienced, "above average pilot-navigator" Flying Officer Denis Briggs of the RAF and co-pilot Ensign Leonard 'Tuck' Smith of the US Navy, which had taken off from Lough Erne, that sighted the battleship *Bismarck*, drawing down the pursuers to gain revenge for the destruction of HMS *Hood*. At the time, the USA being a neutral country, Ensign Smith was one of 17 reported pilots 'on loan' from the US Navy with RAFCC to gain experience of operating this American designed and manufactured flying-boat and amphibious aircraft in combat conditions. He later admitted, "After the original sighting, Briggs turned the controls over to me and went aft to prepare the contact report. The general idea was to take cover in the clouds and close the range somewhat in order to make positive identification. While in the clouds, I misjudged the wind rather badly, and as the record shows, got much too close – right over the ship." Consequently, *Bismarck* opened fire and the Catalina was holed in several places by shrapnel but nevertheless returned to Lough Erne during daylight and managed to alight safely.

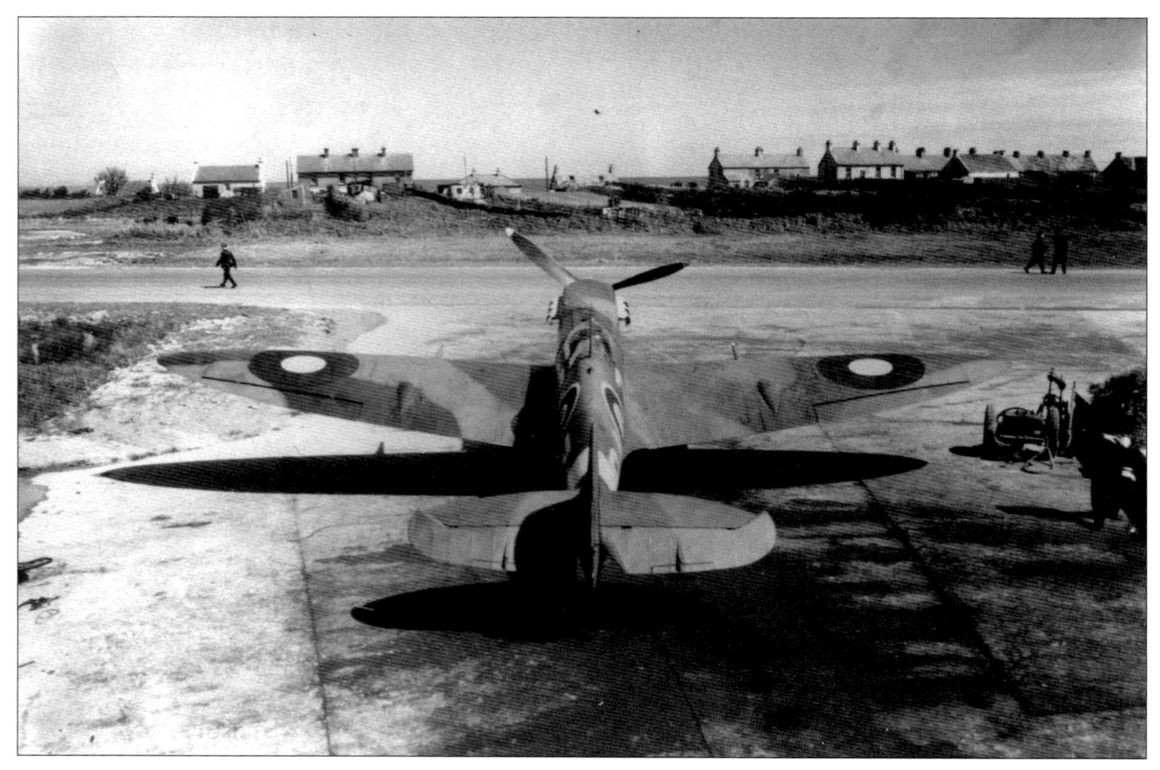

A Supermarine Spitfire of No 130 Squadron at Ballyhalbert. *(Ernie Cromie Collection)*

Ballyhalbert, 20 miles to the south east of Belfast was well placed for the fighter defence of the city. This was Northern Ireland's main fighter station for most of the war, being the base for a succession of squadrons equipped with Hawker Hurricanes, Boulton Paul Defiants, Supermarine Spitfires, Bristol Beaufighters and de Havilland Mosquitos. Aircraft were also detached from time to time to nearby Kirkistown, Ballyhalbert's satellite airfield, to Eglinton in the northwest and St Angelo in Fermanagh. On 23 August 1942, two Spitfires based at Ballyhalbert with No 504 Squadron intercepted a hostile Junkers 88 approaching from south of Dublin and, with some assistance from a lone, Lancashire-based Spitfire which was put out of action by defensive fire, damaged the Ju 88 to such an extent it was forced to ditch in the sea off Waterford. Like many wartime stations, Ballyhalbert was a very cosmopolitan place, one aspect of which was the presence of two Polish squadrons of the RAF, Nos 315 and 303 which were based there from July until November 1943 and November 1943 until April 1944, respectively. The Poles were courageous and high-spirited men, utterly dedicated to the defeat of Nazi Germany to the point of foolhardiness. War graves in Ballycranbeg RC Cemetery and Movilla Cemetery, Newtownards, which contain the remains of Polish pilots who were killed in flying accidents here, are significant testimony to that.

A group of No 133 Eagle Squadron pilots. *(Ernie Cromie Collection)*

The home of the 52nd Fighter Group was at Maydown in 1942. *(Ernie Cromie Collection)*

Prior to the entry into the war of the United States, many American pilots defied their country's policy of neutrality and volunteered to serve in the RAF and RCAF in such numbers that three 'Eagle' squadrons of the RAF were formed in Fighter Command to accommodate them, Nos 71, 121 and 133, which last of the trio was formed at Coltishall in Norfolk in August 1941. In October 1941, No 133 Squadron was sent to the new airfield at Eglinton, with Hurricane aircraft, the first RAF squadron to be based there. Their role was to defend Londonderry and escort convoys in coastal waters. Before returning to Britain at the end of 1941 they were re-equipped with Spitfires, one of which was abandoned over Donegal on becoming lost and out of fuel when returning from patrol. Its occupant, Pilot Officer Roland 'Bud' Wolfe from Nebraska, baled out, was interned in the Curragh but subsequently broke parole and made his way back to Eglinton, only to be handed back to the Éire government by order of Fighter Command's Air Officer Commanding in Northern Ireland who took a dim view of his conduct! Remnants of Wolfe's Spitfire were recovered from a bog in 2011. Several Spitfire squadrons of the RAF also flew from Eglinton and its satellite, Maydown, as did Spitfires of the 52nd Fighter Group of the American 8th Air Force from July 1942, working up to operational readiness prior to their transfer to the 12th Air Force in North Africa in September 1942.

The men and women of No 8 Ferry Pool at Sydenham. *(Ulster Aviation Society)*

A significant event in March 1941 was the establishment of No 8 Ferry Pilots Pool (redesignated No 8 Ferry Pool in May 1942) of the Air Transport Auxiliary (ATA). The CO was OE Armstrong, a very well known pre-war airline pilot, indeed it was he who operated the very first Aer Lingus service from Baldonnel to Bristol in the DH 84, EI-ABI, *Iolar* on 27 May 1936. He is seated fifth from the left in the front row. The prime reason for No 8 Pool's existence was the delivery of Short Stirlings from the manufacturer to the RAF but many other types were flown, including Vickers Wellington bombers to and from Aldergrove and single-engine fighters from the mainland on behalf of the Fleet Air Arm for loading onto aircraft carriers. The Ferry Pool brought the first contact between a young pilot, the future Short's test pilot Tom Brooke-Smith, and Sydenham, a location with which he was to become very familiar over the next 20 years. Tom Brooke-Smith will feature strongly in another volume in this series on aircraft manufacturing in Northern Ireland.

Overhead view of HMS *Gadwall* (Sydenham) Belfast. *(Ernie Cromie Collection)*

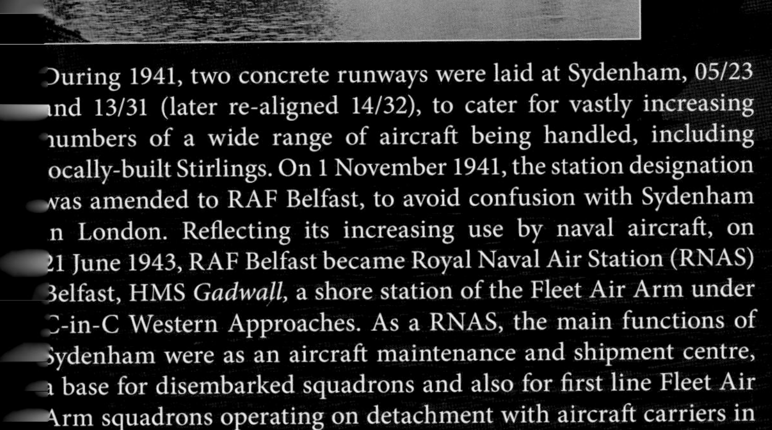

An aircraft carrier and accompanying tug boat at Aircraft Wharf in Belfast Harbour. *(PRONI CAB/3/G/20)*

During 1941, two concrete runways were laid at Sydenham, 05/23 and 13/31 (later re-aligned 14/32), to cater for vastly increasing numbers of a wide range of aircraft being handled, including locally-built Stirlings. On 1 November 1941, the station designation was amended to RAF Belfast, to avoid confusion with Sydenham in London. Reflecting its increasing use by naval aircraft, on 21 June 1943, RAF Belfast became Royal Naval Air Station (RNAS) Belfast, HMS *Gadwall*, a shore station of the Fleet Air Arm under C-in-C Western Approaches. As a RNAS, the main functions of Sydenham were as an aircraft maintenance and shipment centre, a base for disembarked squadrons and also for first line Fleet Air Arm squadrons operating on detachment with aircraft carriers in the North Atlantic. No less than 25 such squadrons passed through between 1943 and 1945. Further types included Gloster Sea Gladiators, Grumman Wildcats, Hellcats and Avengers, Chance Vought Corsairs, Fairey Fireflies and Barracudas. Aircraft came for the American forces too, among them Lockheed Lightnings, Republic Thunderbolts and North American Mustangs for the USAAF. Following unloading, the American aircraft were either towed by road in disassembled state, as in the case of the Lightnings, or assembled on the airfield by personnel of the Lockheed Overseas Corporation and test flown before air-delivery to Langford Lodge to bring them up to operational readiness. As the war drew to a close, Sydenham became chiefly an aircraft storage unit but in 1945 work was begun to develop the site as a largely civilian-manned Royal Naval Aircraft Maintenance Yard for the very specialised task of repairing and reconditioning naval aircraft. Hangars and workshops were erected on the eastern side of the airfield.

A Curtiss Tomahawk of No 231 Squadron, seen here at Sydenham. *(Ernie Cromie Collection)*

Members of the Newtownards Home Guard take aim at a flight of Lysanders. *(Ernie Cromie Collection)*

As one of the Province's two pre-war civil aerodromes, Newtownards was requisitioned after the outbreak of war and used during 1939/40 as a satellite landing ground for Tiger Moths and Miles Magisters from No 24 E&RFTS at Sydenham undertaking practice landings 'circuits and bumps'. It was then decided that Newtownards would be suitable for Army Co-operation flying, so the RAF's No 416 Flight, equipped with Westland Lysanders, was moved from Aldergrove in July 1940 and re-designated No 231 Squadron. Its duties included giving training to anti-aircraft gunners and troops in ground firing at practice targets, aerial photography, patrols over Army road convoys, general support of exercises in various parts of the country, as well as dropping and lifting messages. The following year it received Miles Master IIs to help convert aircrew to a new, much faster American fighter, the Curtiss Tomahawk. In December 1941, the airfield came within the ambit of RAF Fighter Command, reflecting a change of role to accommodate target-towing types such as Lysanders, Masters and Defiants to facilitate air-to-air gunnery practice by fighter types based at Ballyhalbert.

Long Kesh in 1942. *(Ernie Cromie Collection)*

No 231 Squadron moved to the new airfield at Long Kesh in late 1941, joining the Bristol Blenheims of No 226 Squadron, which also flew Army co-operation sorties. With their replacement by arrival of the Douglas Bostons of No 88 Squadron in early 1942, No 231 moved again to another new airfield at Maghaberry. While on the subject of Long Kesh and Maghaberry, it is worth reflecting on the reason for the creation of these two airfields and a relief landing ground at nearby Blaris. As was the case in England, in the early stages of the war, a German invasion of Northern Ireland was a real threat, either by the dropping of paratroops to capture Aldergrove and territory in south Antrim or via neutral Éire; consequently, it was decided to construct the Lisburn airfields to accommodate the air component of British counter-invasion forces. It is interesting to note that, in later years, Long Kesh and Maghaberry became better-known as HM Prisons. Since 2006, one of the historic hangars at the Maze/Long Kesh site has been the home of the Ulster Aviation Society's Heritage Collection of aircraft and much more, commemorating and celebrating Northern Ireland's aviation history.

The Airspeed AS51 Horsa Glider could carry up to 30 troops, a jeep or a 6-pounder anti-tank gun. *(Museum of Army Flying)*

No 651 Squadron Taylorcraft Plus C HH982. *(via Malcolm Coombs)*

Another Army-associated unit, which was based at Long Kesh for a time, was 'A' Flight of No 651 Air Observation Post (AOP) Squadron, which was equipped with recently commandeered civilian Taylorcraft (the forerunner of the famous Auster). It is also of interest to note in this connection that, in 1942, Long Kesh, as well as Nutts Corner, became the Northern Ireland terminus for Army aviators of the Glider Pilot Regiment who participated in a towed-glider service from Netheravon in Wiltshire, the purpose of which was to give glider crews and troops of the Airborne Division training in long distance navigation and to test airborne equipment before the great glider-borne assaults of later years on Hitler's 'Fortress Europe'. Usually, the aircraft used were Stirling and Whitley bombers towing Hotspur or Horsa gliders and the sorties were not without incident. On 11 August 1942, for instance, a Hotspur being towed by a Whitley had to be abandoned in bad weather, the glider being damaged beyond repair when it forced landed on the decoy site near Stoneyford, happily without loss of life.

Wildcat V at Eglinton in 1944. *(Mick Burrow)*

The Escort Carrier HMS *Searcher*. *(Raymond Burrows Collection)*

In 1943 Eglinton was transferred to the Royal Navy, along with its satellite airfield at Maydown. The first RN aircraft to arrive were the Fairey Swordfish of 835 and 837 NAS from Ballykelly. HMS *Gannet*, as it became, was used primarily for the working up of RN fighter units – Wildcat, Martlet, Seafire, Hellcat and Corsair. Three Naval Fighter Wings (NFW) were formed there, the 5th, 7th and 10th. Squadrons formed, re-formed and trained prior to embarking either onto large Fleet Aircraft Carriers such as HMS *Indefatigable* or HMS *Glory*, or the smaller Escort Carriers such as HMS *Searcher* or HMS *Pursuer*.

Hellcats at Eglinton in 1944. *(Mick Burrow)*

Corsair KS774 of 1851 NAS at Eglinton in January 1945. *(Raymond Burrows Collection)*

The Escort Carrier concept was of vital importance in winning the Battle of the Atlantic. These vessels were essentially mass-produced hulls, mostly constructed speedily in US shipyards and fitted with a flight deck, arrester wires and a barrier. The Royal Navy received 35 out of the total of 133 constructed for use by the Allies, the key to their success being the ability to provide air cover at sea where it was most needed, directly over the convoys. The first two Grumman F6F Hellcat squadrons to commission for operational FAA service, 800 NAS and 804 NAS, did so from Eglinton in 1943, from whence they proceeded to the escort carrier, HMS *Emperor,* for service on the convoy run to Gibraltar initially. The American-built Grumman Wildcat, Grumman Hellcat and Chance Vought F4U Corsair were the supreme carrier-borne fighter aircraft of the Second World War. The Wildcat and Hellcat combined great structural strength and excellent manoeuvrability, the Hellcat adding much improved performance. The Corsair achieved particular fame in the hands of the pilots of the US Marine Corps fighting in the Pacific.

A striking view of the MAC ship *Macoma* taken from a departing Swordfish. *(Royal Netherlands Navy)*

Maydown was the headquarters base for MAC ship (Merchant Aircraft Carrier) operations. This third type of aircraft carrier proved to be a highly effective countermeasure to the U-boat offensive from mid-1943 onwards. These were standard grain carriers or oil tankers fitted with an elementary flight deck from which a Flight of three or four Swordfish was operated. Each Flight of Swordfish flew from Maydown to join the carrier off the Irish coast and returned to base after the journey across the Atlantic and back. It is a remarkable fact that, of the 217 convoys in which a MAC ship sailed between May 1943 and the end of the war, only one was attacked successfully by a U-boat. Normally two MAC ships would escort a convoy. Three parent units for the Swordfish were based at Maydown, 836 and 860 NAS for operational deployment, and 744 NAS for training. 836 was the largest operational squadron in the FAA. Together the Maydown squadrons provided over 90 Swordfish for some 19 MAC ships. The last FAA squadron to relinquish the famous Swordfish was 836 at Maydown in July 1945.

Sub-Lieutenant (A) Peter Lock. *(Peter Lock)*

Grumman Wildcat Mk V, JV705, of 882 NAS in November 1944. *(via John McMullen)*

FAA types also visited Long Kesh for Army Co-operation and other training. Some were based there temporarily, including 882 NAS in late 1944. On Christmas Eve 1944 the Grumman Wildcat Mk V, JV482, took off from Long Kesh with another Wildcat from the squadron to carry out some target practice at Lough Neagh but very shortly afterwards had to ditch in shallow water in Portmore Lough because of an engine fire. The pilot, Sub Lieutenant (A) Peter Lock RNVR, was safely brought to shore by a local boatman and did not even get his feet wet! Transport had been sent from Long Kesh and soon he was back in the mess enjoying his Christmas Eve dinner. His aircraft remained partially submerged for 40 years. It was eventually recovered in 1984 and now is a major restoration project in the Ulster Aviation Collection at the Maze/Long Kesh site.

Ballyhalbert in late 1944. *(Ernie Cromie Collection)*

On 24 April 1945, Ballyhalbert was transferred to the Admiralty and commissioned as HMS *Corncrake*, but its control by Their Lordships was short-lived for, on 13 November 1945, despite having become the base of No 4 Naval Air Fighting School, it was paid off and returned to the RAF. This was very ironic indeed in light of the fact that since October 1943, when the first one arrived, many more squadrons of the Fleet Air Arm, first and second line, had 'lodged' at Ballyhalbert, for rest and training purposes, than resident RAF units. Their equipment included a wide range of aircraft – Seafires, Wildcats, Hellcats, Corsairs, Fulmars, Martinets and Barracudas. One of the squadrons, No 885, which was there for just over four months, had been in action with Seafires over Normandy during D-Day and its aftermath but it took the opportunity at Ballyhalbert to re-equip with Hellcats. One of its pilots was the late George Boyd, whose family home was at Killinchy on the opposite side of Strangford Lough and whose memoirs were published by Colourpoint Books in 2002 as *Boyd's War*. Ballyhalbert's satellite airfield at Kirkistown also became navalised as HMS *Corncrake II*.

US technicians at Killadeas, Co Fermanagh in August 1941. *(Clive Moore Collection)*

Civilian workers from the United States were at work in Northern Ireland before America entered the war in December 1941. They were engaged in construction work in connection with improving the dockyard in Londonderry and in the creation of a base for US Navy Catalina flying-boats on Lough Erne, a facility that the organisation decided it did not require and instead it became RAF Killadeas. From 26 January 1942, when the first US troops to arrive in the British Isles disembarked in Belfast, until after the D-Day landings, Northern Ireland played host to many thousands of US soldiers, sailors, airmen and civilians in an environment which was far enough away from the mainland of Occupied Europe to be free of sustained air attack, apart from the 'Blitz' of Spring 1941.

Nutts Corner in 1945. *(Ernie Cromie Collection)*

One aspect of the American presence in the UK during the Second World War was the mass ferrying of hundreds of bombers and large transport aircraft by air across the Atlantic, under the auspices of the USAAF's Air Transport Command. The corresponding organisation on the British side was RAF Ferry Command, the foundation and inspiration of which was the delivery, on 11 November 1940, of a group of seven Lockheed Hudsons for the RAF, from Botwood in Newfoundland to Aldergrove. The leader of the formation was the renowned Captain Donald Bennett who later created 'The Pathfinders'. To facilitate this huge operation, four airfields were re-developed as Transatlantic Ferry Terminals, three in Britain and a fourth at Nutts Corner, formerly an RAF Coastal Command base. Although all four remained under RAF control, each, significantly, was designated an American Army Air Force Station, Nutts Corner being AAF Station 235. USAAF aircraft arrivals at Nutts Corner were recorded in gradually increasing numbers from July 1943 and a year later the 1404th AAF Base Unit was established there, followed soon afterwards by a station to obtain essential data for weather forecasting. July 1944 was Nutts Corner's record-breaking month when 372 aircraft arrived: 246 B17s, 90 B24s, 12 B26s and 24 C47s. In 1945 the airfield came under Royal Navy jurisdiction as HMS *Pintail*. It was equipped to handle first line fighter squadrons with a capacity for 60 aircraft but with the end of the war at hand it was used principally for the disbandment of Seafire and Corsair squadrons. The extensive aircraft parking areas constructed to facilitate the airfield's use as a Ferry Terminal provided a good basis for its choice as the site of Belfast Civil Airport in 1946.

Toome crew training room. *(Ernie Cromie Collection)*

USAAF officer with RAF officer at Toome with a B17G in the background. *(PRONI CAB/3/G/13/6)*

Toome, Station 236, in May 1944 from a water colour painting by Doris Blair.

The Lockheed Overseas Corporation Chief Test Pilot, George Clark, inspects a B17 at Langford Lodge. Note the B24 by the hangar in the background. *(Ernie Cromie Collection)*

Plans drawn up shortly after United States' entry into the war envisaged Northern Ireland becoming the focus of an extensive programme of aircrew training for the European Theatre of Operations (ETO). In 1943, Combat Crew Replacement Centres were activated at Toome, Cluntoe and Greencastle for the training of crews for B26 Marauders/A20 Havocs and B17 Flying Fortresses respectively. Cluntoe/Greencastle changed to B24 Liberators in 1944. In the case of the heavy bombers (B17s and B24s), the officer element of the crews, normally the pilots, co-pilots, navigators, bombardiers and radio operators received their training at Cluntoe while the gunners went to Greencastle where air-to-air and ground-to-air gunnery schools and a tow-target flight was based. Greencastle also served as a satellite air depot for Langford Lodge, where there was insufficient room to accommodate all aspects of the logistical effort there, indeed, at times facilities were almost overwhelmed by the sheer numbers of aircraft requiring attention. Maghaberry was used primarily as a base for aircraft ferry squadrons and also as a transit point for casualty evacuation, due to the proximity of the US Army's 79th Station Hospital at Moira. Long Kesh was a transit stop for USNAS communications flights operating between RAF Hendon in London and Eglinton near Londonderry.

US aircraft at Langford Lodge. *(Ernie Cromie Collection)*

An aerial view of Langford Lodge in 1945. *(Ernie Cromie Collection)*

Langford Lodge became one of four primary air depots for the USAAF in the United Kingdom, which reassembled, serviced, overhauled, repaired and salvaged US aircraft for the European and Mediterranean Theatres of Operations. Uniquely, it was operated contractually for the first two years of its life by the Lockheed Overseas Corporation (LOC). Thousands of American bomber, transport and fighter aircraft were modified for active service at Langford Lodge. In its heyday, in 1943–44, some 7500 people were employed there at the 3rd Base Air Depot – comprising about 1500 military personnel, about 3000 Lockheed staff and a similar number of local civilians, paid for from Lockheed funds. A truly staggering total of 3250 new aircraft were assembled there, more than 11,000 were serviced and 450,000 components were overhauled. The scale of activity is demonstrated by the accompanying photos which were taken around the end of the war when large numbers of aircraft were present at a particular point in time. For instance, records show that on 17 May 1945, 572 aircraft were parked on and around the airfield, most of them war weary aircraft to be salvaged. To relieve the pressures of essential logistical work, famous entertainers and personalities who visited the base included Al Jolson, Bob Hope, Frances Langford, Bebe Daniels, Glenn Miller and Joe Louis.

Consolidated PB2Y-3R Coronado – sadly not over Northern Ireland. *(US Navy)*

Sandy Bay. *(Ernie Cromie Collection)*

The preparations for D-Day intensified in early 1944. Essential personnel and urgent stores were flown by US Naval Air Transport Service Consolidated PB2Y-3R Coronado flying-boats on a daily basis, on a scheduled service from New York via Botwood in Newfoundland to Sandy Bay on Lough Neagh, flown by Pan American Airways and American Export Airlines crews. The first to arrive was BuNo 7219, flown by Captain Durst, on the morning of 18 May 1944. This was the first of 538 Atlantic crossings made by the PB3Y-3Rs during that summer. In June the route was extended to Port Lyautey in North Africa. As many as 11 flying-boat movements per day were recorded in the run up to D-Day. Sixty arrivals were recorded in June, carrying 280 passengers. The normal load was between 10 and 18 passengers, and 9 crew. The sole accident for the type on Lough Neagh was on 17 July when BuNo 7223 struck some rocks and sustained damage. It was repaired after being raised and towed ashore by the McGarrys and returned to service. On 21 July, Captain Olaf Abrahamson of Pan American took off from Sandy Bay in BuNo 7230 and arrived at Shediac, New Brunswick in the record time of 14 hours 18 minutes. The service terminated on 15 October 1944

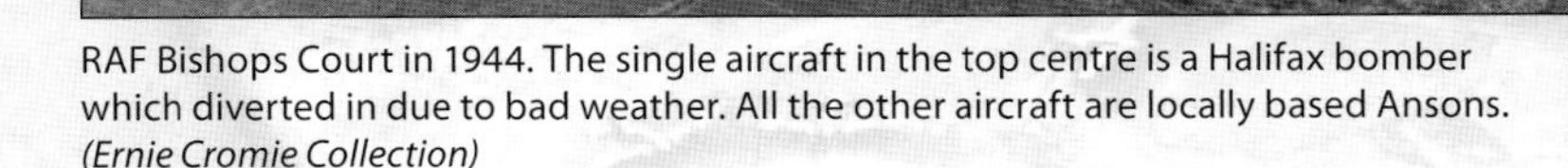
RAF Bishops Court in 1944. The single aircraft in the top centre is a Halifax bomber which diverted in due to bad weather. All the other aircraft are locally based Ansons. *(Ernie Cromie Collection)*

General Eisenhower with VIPs at Long Kesh in 1945. *(PRONI CAB/3/G/7)*

Many ships of the vast D-Day invasion fleet assembled in Belfast Lough. As events moved rapidly in France so also in the Province. By the end of 1944 only two of the six airfields which had been handed over to the USAAF remained in American hands, Langford Lodge and Greencastle. On 23 August 1945, General Eisenhower flew into RAF Bishops Court for a one-day visit to receive the Freedom of Belfast and pay warm tribute to Northern Ireland's wartime role. After commenting, "Without Northern Ireland I do not know how the American Forces could have concentrated to begin the invasion of Europe", he left on the following day from RAF Long Kesh. In March 1946, the US flag came down at Langford Lodge, bringing to an end a remarkable feat of organisation and training.

One of the new airfields was the Combat Crew Replacement Centre, Greencastle. *(Ernie Cromie Collection)*

RAF Maghaberry. *(Ernie Cromie Collection)*

More than 20 new airfields were constructed in Northern Ireland during the war. Some performed an unglamorous but none the less vital role as Satellite Landing Grounds (SLGs). Essentially they were outstations of No 23 MU at Aldergrove and were used to store the vast reserve of aircraft upon which the RAF could call as production intensified during the war. Chief among these were Murlough and Ballywalter. Aircraft stored included Ansons, Blenheims, Wellingtons, Lysanders, Hampdens, Herefords, Whitleys, Battles, Bothas and Corsairs. As a final sad footnote to the story of the SLGs, after the war many hundreds of aircraft were broken up for disposal on the storage sites, including some 200 Stirlings at Maghaberry alone – if only just one example had been preserved! It is also of interest to note a very unusual difficulty that had to be overcome by the engineers who constructed the airfields – the reluctance of local people to remove fairy thorn bushes from the middle of proposed runway sites.

Spitfire Mk VII MD159 of the Met Flight at Aldergrove. *(I Henderson Collection)*

During the war, the Met Flight which had been established at Aldergrove in 1936 expanded significantly. In December 1939, its Gauntlet aircraft were replaced by Gladiators, which remained on strength until February 1945. In January 1941 it was redesignated 1402 Flight and two months later 1405 Flight was formed to operate alongside it with Blenheims, the two Flights being amalgamated in March 1942 following the replacement of Blenheims with longer range Hudsons. This reflected a growing requirement for much longer range aircraft suitable for travelling farther over the Atlantic and large enough to carry sufficient instruments and weather observers to obtain as much scientifically measured data as possible. Other aircraft to serve with the Aldergrove Met Flights included Handley Page Hampdens and Hurricanes, and a few of the rare Spitfire Mk VI and VII high altitude variants were also supplied, which were capable of ascending to 43,000 feet. They were the only Spitfires to be based at Aldergrove during the war. It was September 1945 before four-engined 'Met' Halifax aircraft began to operate from the station, marking the arrival of No 518 Squadron from Tiree.

Blackburn Shark target towing at Armament Training Camp, Lough Neagh in 1940. *(Ernie Cromie Collection)*

A flavour of the aerial activity in and around Aldergrove towards the later stages of the war may be gained from examining the records of the Armament Practice Camp for a typical day – Saturday 22 April 1944. First on the scene in the morning, were the Liberators from No 1674 Heavy Conversion Unit at Aldergrove to drop 24 bombs, they were followed by Wellingtons from No 407 Squadron at Limavady with another 36 bombs. In the afternoon, the Liberators came back again with a further 47 bombs to which could be added a Short Sunderland from No 422 Squadron at Castle Archdale. While all of this multi-engine activity was going on, Supermarine Seafire naval fighters from Long Kesh were busy unloading 89 bombs. But for inclement weather which caused its cancellation, the late evening calm would have been shattered by a Leigh Light bombing exercise carried out by more Liberators from No 120 Squadron, Ballykelly. In all, there were 15 functionally different target zones dispersed throughout Lough Neagh.

A dramatic image of the sinking of U-265 on 10 March 1944 by Sunderland EK591 of No 422 Squadron, Castle Archdale. *(Ernie Cromie Collection)*

The final two U-boat sinkings by Province-based aircraft occurred on 29 and 30 April 1945, by a Liberator of No 120 Squadron and a Sunderland of No 201 Squadron respectively. When No 120 Squadron disbanded at Ballykelly on 4 June 1945 it shared the honour with No 86 Squadron, then based at Tain in Scotland, of being Coastal Command's most successful U-boat slayers, each with 14 confirmed kills. No 201 Squadron was credited with sinking five U-boats. However, it is unfair to compare one squadron with another in this way; they all served, if only to keep U-boats submerged and therefore unable to keep pace with convoys. Moreover, it is advisable to regard statistics about the destruction of enemy submarines with a degree of scepticism, as the precise circumstances relating to the loss of every single one that failed to return from patrol will never be known. It can safely be said, however, that without the efforts of the crews of these two magnificent aircraft, the Short Sunderland and the B24 Liberator, the Battle of the Atlantic would not have been won.

Wing Commander Barrett, his aircrew and ground crew pose beside a Royal Air Force pinnace at Castle Archdale. *(Ernie Cromie Collection)*

The final operational patrol of the war from Castle Archdale was on 3/4 June 1945, carried out in Sunderland 'Z' of No 201 Squadron, captained by Wing Commander John Barrett, DSO and crewed, appropriately, by a composite group of Australians, New Zealanders, Canadians and Britons who escorted an inbound convoy of 51 ships. It was also the last wartime patrol of Royal Air Force Coastal Command. The Command's casualties during the war, due to hostile actions and accidents but excepting death from natural causes, amounted to 8,874 in total and 2,601 wounded, the figures being inclusive of Dominion and Allied personnel.

HM the King arrives at Eglinton. *(FAA Museum)*

On 17 July 1945, Their Majesties King George VI and Queen Elizabeth with their daughter Princess Elizabeth flew into Long Kesh airfield to begin a ceremonial visit to Northern Ireland as part of a 'Victory Tour' of the UK. This was the first occasion on which they visited the country by air, the King's aircraft being No 24 Squadron's highly polished VIP Dakota KN386. The visit ended on 19 July when they flew to RNAS Eglinton, where they were met by local dignitaries and officers of the Royal Navy. In the course of their visit to Londonderry, they went to Lisahally and took the opportunity to cast their eyes over 52 surrendered U-boats tied up there. Two in particular attracted their close attention, U-2513 and U-3008, revolutionary Type XXIs, the ultimate version which, fortunately for the Allies, entered battle too late to have a crucial effect on the outcome of the war.

The RAF Pinnace 1251 at Strangford in 1943. *(Ernie Cromie Collection)*

A Sunderland and tender on Lough Erne. *(Ernie Cromie Collection)*

The Marine Branch, which ceased to exist as a separate entity in 1986, is a comparatively neglected aspect of RAF history, yet in rudimentary form it was inherited when the service was created in April 1918 by the amalgamation of the RFC and the RNAS. The origins of its presence in Northern Ireland are a little obscure but it would appear that RAF marine craft may have been based here initially towards the end of the 1920s, on the eastern shore of Lough Neagh near Ardmore, to assist in the use of the lough by military flying-boats during occasional visits to RAF Aldergrove. Be that as it may, there is no doubt that, during the Second World War, RAF marine tenders were supplied to the civilian McGarry organisation at Ardmore to facilitate their contracted responsibility for maintaining targets and salvaging aircraft which crashed into the lough. In addition, an RAF marine craft unit, operating from Antrim and Sandy Bay, complemented the work of the McGarry enterprise. By 1943, RAF Marine Craft Units were based at Portaferry, Donaghadee, Larne, Londonderry, Lough Erne and at Portrush for air/sea rescue and other purposes.

Greystone radar. *(Ernie Cromie Collection)*

One of the critical assets that enabled RAF Fighter Command to make efficient use of limited numbers of fighter aircraft for defence purposes, certainly in the early stages of the war, was radar. In Northern Ireland, it was 1941 before the first of several types of radar station came into effective use although by the end of 1944 a significant number had been closed, reflecting a reduction in the number of incursions by Luftwaffe aircraft. For the most part, they were peripherally located, the most significant locations being Ballydonaghy (Strabane), Castlerock, Glenarm, Black Head, Greystone, Roddans Port, Ballywoodan, Kilkeel, Ballymartin and Lisnaskea. Some of the associated structures remained in place until well after the war, for instance at Greystone between Millisle and Ballywalter, clearly discernible in this photograph taken on 14 July 1951 are two Chain Home aerial masts, betrayed by the long shadows thrown to the west of them by the early morning sun.

Handley Page Halifax GR6 and Spitfire Mk VII MD159 of No 518 Squadron fly in formation. *(I Henderson Collection)*

A Hastings of No 202 Squadron takes off on a Met flight. *(RAF Aldergrove)*

On 18 September 1945, the Handley Page Halifaxes of No 518 Squadron moved to Aldergrove from the Scottish island of Tiree. Their task was meteorological and as a large unit the squadron absorbed the existing No 1402 Flight with Spitfire VIIs and Hurricane IICs, which had returned to Aldergrove after a few months at Ballyhalbert. By the middle of 1946, No 518 Squadron at Aldergrove was the only Met unit in the UK. In October, it was renumbered as No 202 Squadron. The Halifaxes soldiered on until October 1950 from which month they began to be replaced. In little over five years of operations from Aldergrove, several of the aircraft were lost while engaged on Met duties. They were replaced by another Handley Page product, the Hastings, which took its first operational sortie on 14 December 1950. The Hastings was easier and more comfortable to operate than the Halifax; moreover, the type had an excellent safety record compared to the Halifax, accidents in which resulted in the deaths of 32 crew members. The 'Met' Hastings era and that of RAF Meteorological Squadrons came to an end with the disbandment of No 202 Squadron at Aldergrove on 31 July 1964.

A Gloster Javelin delta-wing fighter seen here at Aldergrove with personnel from No 23 MU. *(Ernie Cromie Collection)*

Ecuadorian Air Force Canberra recovered by No 23 MU. *(Ernie Cromie Collection)*

The end of the war also brought changes for No 23 MU. Its first task was in sharp contrast to the restorative work it had carried out on many hundreds of aircraft – the storage and scrapping of aeroplanes which were now surplus to requirements. However, there were still other aircraft, which needed repairs, maintenance and modification. Many of the new aircraft joining the RAF inventory made their way to Aldergrove in the late 1940s. No 23 MU was kept very busy during the 1950s with Lincolns, Sabres, Shackletons, Swifts, Javelins, Hunters and Varsities. It also had the much sadder job of scrapping numerous surplus aircraft, including several Boeing Washington bombers – B29 Superfortresses, the same type that had dropped the atomic bomb – supplied to the RAF during the Cold War. In 1957, No 23 MU absorbed No 278 MU, which had been formed at Aldergrove nine years earlier and in August 1962, personnel from both were involved in a most unusual achievement, the dismantling and recovery by road of a Canberra B6 bomber of the Ecuadorian Air Force which had crash-landed in the Republic of Ireland. The dismantled airframe was shipped to Britain where it was repaired and returned to Ecuador.

No 502 Squadron received its first jet aircraft in 1951, DH Vampires. *(Ernie Cromie Collection)*

A No 502 Squadron group in front of a Mosquito in 1947. Squadron Leader McGiffin is in civilian dress and is standing eighth from the left. *(via Jack Greer)*

When No 502 Squadron Royal Auxiliary Air Force (RAuxAF) was re-embodied on 17 July 1946, with Ulsterman and former member Hunter McGiffin in command in the rank of Squadron Leader, there was no shortage of volunteers for 'weekend flyer' duties, despite the fact that recruitment was initially restricted to former RAF personnel. Initially a light bomber unit, it operated de Havilland Mosquito B25s and NF30s before being re-equipped with Spitfire F22s on becoming a day fighter unit, a new experience for squadron members, which they took to enthusiastically and expertly, managing to win the Cooper Trophy in 1949. During air firing sessions at their last annual summer camp in Yorkshire in 1956, they proved to be not only the leading RAuxAF squadron in the UK, their results were superior to the two regular RAF units on the station. By then, they were operating de Havilland Vampire FB5, FB9 and T11 jets, which had begun to replace the Spitfires in 1951.

Firefly FR1 MB726 of 814 NAS over Eglinton in 1947. *(Don MacGregor)*

A Fairey Barracudda III of 744 NAS, Maydown in 1946. *(John Devenney via Raymond Burrows Collection)*

After the war, Eglinton was retained as a RNAS firstly with fighter aircraft, Wildcats, Corsairs, Fairey Fireflies, Hawker Sea Furies and Supermarine Seafires, and then with anti-submarine types – Fairey Barracudas, Firefly AS6s and Grumman Avengers. In the early 1950s it was the Naval Air Anti-Submarine School and from 1955 onwards became especially associated with the Fairey Gannet AS1s and T2s of 719 and 737 NASs. In November 1957, 719 absorbed 737 and the name was changed to the Naval Anti-Submarine Operational Flying School. Additionally the Gannets of 812, 815, 820 and 847 NASs formed at Eglinton in 1955 and 1956 before embarking on the carriers HMS *Eagle*, HMS *Ark Royal*, HMS *Centaur* and HMS *Bulwark*. Moreover the Joint Anti-Submarine School (JASS) was situated at HMS *Sea Eagle* in Londonderry.

An advert placed in *Flight* magazine by English Electric Ltd, celebrating the Canberra's record breaking achievement. *(via Paul McMaster)*

The Trans-Atlantic double record breaker Canberra B5 prototype VX185. *(Ernie Cromie Collection)*

Famous as the first jet bomber to be produced in the UK and the first to serve with the RAF, the English Electric Canberra first flew at Warton in Lancashire on 13 May 1949. It was a success from the start and numerous variants were developed and constructed to satisfy the requirements of at least 16 different countries throughout the world, including the UK and USA, work being sub-contracted to various companies including Shorts in Belfast, the first production Canberra from which flew in October 1952. Meanwhile, Northern Ireland had first featured in the Canberra story when, on 21 February 1951, the B2 example WD932 took off from Aldergrove and landed at Gander, Newfoundland only 4 hours and 37 minutes later, a new record time for the Atlantic crossing. On 31 August 1951, that record was broken by another B2, WD940, which took only 4 hours and 18 minutes to make the same journey. More remarkably still, on 26 August 1952, the prototype B5 Canberra was flown from Aldergrove to Gander and back again in 10 hours and 3 minutes, including a refuelling stop, the actual flying time for the return trip being 7 hours and 58 minutes.

This Whirlwind Mk 7 from 820 NAS made a forced landing near the Giant's Causeway on 12 March 1958. *(Hugh McGrattan)*

Gannets of 824 NAS on the dispersal area at Eglinton in 1958. *(FAA Museum)*

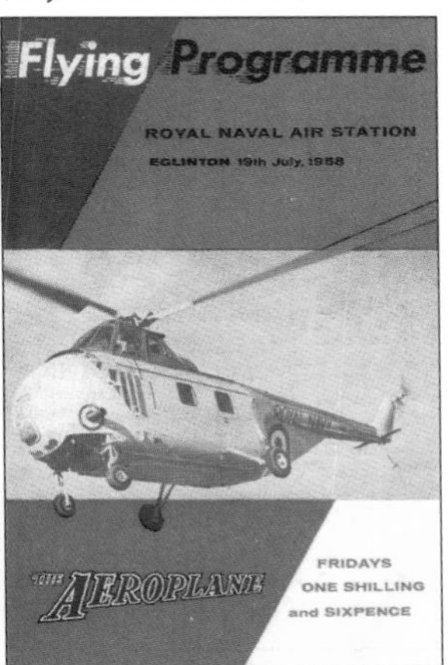

Eglinton Open Day 1958. *(Guy Warner Collection)*

Westland Dragonfly 917. *(Raymond Burrows Collection)*

Avro Shacketon MR1a VP291 over the rugged north Antrim coast. *(David Hill)*

Post-war, Avro Shackletons were based at Ballykelly from 1952 onwards and Short Sunderlands flew to Lough Erne every year to take part in JASS courses. The Sunderlands retired from RAF service in 1957 and the Gannets left Eglinton in 1959, though it was still used by RN helicopters – Dragonfly, Whirlwind and Wessex – until 1966. The Shackletons carried on at Ballykelly until 1971. Aldergrove became a maritime reconnaissance base again on 1 April 1952, with the arrival of No 120 Squadron, which was equipped with the Avro Shackleton MR1.

Fire fighting drill at Sydenham. *(Guy Warner Collection)*

A formation flypast of four Sea Hawks at a Sydenham Air Pageant. *(via Liz Shanks)*

Military activity also continued at Sydenham in the 1950s. Throughout the period, Queen's University Air Squadron was a focal point for flying training, receiving the long serving de Havilland Chipmunks in 1950, two North American Harvards were on the establishment in the mid 1950s, while the Hunting Percival Provost appeared for a time in 1956. Royal Naval Aircraft Yard (RNAY) Sydenham still continued in being, with maintenance and repair work being carried out on a new generation of naval aircraft including Sturgeons, Sea Venoms, Meteors, Sea Hawks, Hunters and Sea Vixens. The general public was entertained by the annual Air Pageants organised by Royal Air Force Association (RAFA) between 1947 and 1956. These were always well-attended and featured many star attractions of the day.

Photographs of aircraft at Bishops Court are rare. Here is one of a Vickers Wellington T Mk X of the Air Navigation School – the date is not known but is probably in the mid-1940s. *(RAF Museum)*

Large parade at RAF Bishops Court. *(PRONI CAB/3/G/18)*

The outbreak of the Korean War in June 1950 caused a re-thinking of RAF training requirements which, with regard to Air Navigation Schools (ANS), resulted in two comparatively short-lived developments in Northern Ireland. No 4/5 ANS functioned to a limited degree at a partially re-furbished RAF Langford Lodge from September 1952 until the end of January 1953, while No 3 ANS fared somewhat better at RAF Bishops Court, forming there at the beginning of March 1952 and being disbanded in mid-April 1954, having been equipped with Vickers Valettas, Varsities and Avro Ansons. It was also planned that Bishops Court would become an all-weather fighter base and emergency dispersal airfield for the V-bomber force but, whereas some modifications were made to the runways and aircraft dispersals, defence cuts caused this proposal to be abandoned. Enthusiastic aircraft spotters, eagerly anticipating Hunters, Javelins, Valiants, Vulcans and Victors had instead to content themselves with a few Kirby Cadet and Slingsby Sedbergh gliders which arrived in 1959 with the RAFVR's No 671 (Volunteer) Gliding School, which was based there until 1962, giving Air Cadets gliding instruction, of which more later. Bishops Court eventually closed in 1990, with the removal of a facility housing a mobile Air Defence Radar.

The RAF Aldergrove Freedom of Belfast ceremony in 1957 in front of the City Hall. *(Ernie Cromie Collection)*

Presentation of 502 Squadron Standard by Lord Wakehurst. *(Ernie Cromie Collection)*

Sadly No 502 Squadron, along with all the other Royal Auxiliary Air Force squadrons was disbanded in 1957, after a Government Defence Review. It was a sad but proud body of men who paraded at Aldergrove on 10 February 1957 to witness the squadron's last operational flight, a flypast by three Vampires. No 502 Squadron's record of wartime service must never be forgotten – five U-boats and a huge tonnage of ships destroyed, and a total of 174 men officially listed as killed or missing, the last resting places of 107 of them being unknown. The Squadron Standard, emblazoned with five Battle Honours, hangs in St Anne's Cathedral, Belfast as a poignant reminder of the unit's unique and sacrificial contribution to an important aspect of UK military aviation history and the defeat of Fascism. It was presented during a moving ceremony at RAF Aldergrove on 24 May 1954, by His Excellency the Governor of Northern Ireland, Lord Wakehurst, KCMG who was accompanied by the Honorary Air Commodore of the Squadron, Viscount Brookeborough of Colebrooke, CBE, MC, DL, MP and Prime Minister of Northern Ireland. On 5 October of the same year RAF Aldergrove was granted the Freedom of Belfast. For a time in the late 1950s the Hawker Hunters of Nos 19 and 65 Squadrons and the Gloster Meteor NF14s of No 153 Squadron were detached to Aldergrove on rotation to provide fighter cover of the western extremity of NATO during the Cold War.

13 Flight Auster AOP6. *(Raymond Burrows Collection)*

1913 Light Liaison Flight at Aldergrove in 1957. *(Captain Peter Wilson)*

Saro Skeeter XN350 at Newtownards. *(Tommy Maddock)*

On 14 February 1957 new aircraft arrived at RAF Aldergrove in the shape of the five Auster Mk 6s of 1913 Light Liaison Flight. They soon became 13 Flight 651 Squadron, of the newly revived Army Air Corps (AAC). To begin with the flight had been under the control of the RAF, however the pilots were all members of the Glider Pilot Regiment. For a period a small detachment was based at St Angelo. The Austers were soon joined at Aldergrove by Saunders Roe Skeeters. 13 Flight remained at Aldergrove until November 1962, when it was replaced by 2 Reconnaissance Flight, 2 Royal Tank Regiment, with similar aircraft.

A fine study of Shackletons of No 203 Squadron in formation. *(via David Hill)*

Shackleton MR3s of No 120 Squadron. *(RAF Aldergrove)*

No 120 Squadron received its first Shackleton MR3 in July 1958, replacing the MR2s, which had been on strength for the previous two years. These improved versions of a much loved and reliable aircraft were held in high regard by the crews – though placing mats at the entrances, on which they were expected to wipe their feet before boarding, does seem a little excessive. On 1 April 1959, seven years to the day after arriving at Aldergrove, the Shackletons departed for the last time to their new base at Kinloss.

A No 118 Squadron Sycamore with two RUC personnel on board. *(Group Captain David Toon)*

A No 118 Squadron Sycamore over the eastern edge of Lough Neagh. *(Group Captain David Toon)*

The late 1950s saw an upsurge of IRA bombing and shooting incidents along the border of Northern Ireland and the Irish Republic. This became known as the Border campaign and it lasted from December 1956 until February 1962. In all there were about 600 incidents of violence which resulted in the deaths of six policemen and ten IRA volunteers. On 1 September 1959, after a spate of bombings, No 118 Squadron Royal Air Force was reformed at short notice at Aldergrove from a detached flight of No 228 Squadron – a Search and Rescue (SAR) unit. The squadron's Sycamores proved very effective in co-operation with the RUC and many members of the full-time and reserve forces flew on cordon and search or reconnaissance missions between 1959 and 1962. Previously F Flight of No 275 Squadron had introduced the Sycamore to the Province, when based at Aldergrove in the SAR role between 1957 and 1959. On 31 August 1962, No 118 Squadron disbanded.

Aldergrove 'At Home' day brochure 1960. *(Guy Warner Collection)*

A Hastings of No 202 Squadron on static display at the Aldergrove 'At Home' Day in 1960. Note the formation of Gloster Javelins above and behind. *(Tommy Maddock)*

In 1960 Aldergrove was still functioning as the main RAF base but with only the Handley Page Hastings of No 202 Squadron flying weather sorties, the Bristol Sycamores of No 118 Squadron and the Austers and Skeeters of 13 Flight AAC in residence, along with No 23 MU. Shackletons of Nos 203, 204 and 210 Squadrons flew maritime patrols from Ballykelly, a radar station scanned the skies from Bishops Court, while naval helicopters were based at Eglinton. Belfast Harbour Airport at Sydenham was the Short Brothers company airfield and also a Royal Naval Aircraft Yard for an assortment of aircraft in storage, including Sea Hawks, Sea Venoms and Sea Vixens.

A Bristol Belvedere brings supplies to a farmer. *(Newsletter 13 February 1963)*

Two Skeeters delivering hay in the Mournes. *(Newsletter 10 January 1963)*

The Skeeter which crashed at Mossley in 1963. *(Guy Warner Collection)*

A Skeeter lands in Castlewellan in 1963. *(Guy Warner Collection)*

Wessex HAS1 XM916 of 819 NAS assists with milk deliveries to Coleraine in February 1963. *(Hugh McGrattan)*

Wessex HAS3 XP118 at Ballykelly. *(Hugh McGrattan)*

Bringing much-needed and appreciated aid of a humanitarian kind to the civilian population has been a long-standing feature of the history of military aviation here, as elsewhere. The very harsh winter of 1962/63 is a case in point. Exceptionally cold temperatures set in before the New Year with freezing conditions being recorded in January 1963 over a consecutive period of days, to such an extent that comparison was made with the two previous worst winters on record, those of 1947 and 1895. With the frost came heavy falls of snow and many roads became impassable, with some towns and villages cut off by head-high snowdrifts. Following a request for help from the Minister of Home Affairs, Brian Faulkner, aircraft of all three services, especially helicopters, rendered considerable assistance to the public across Northern Ireland over the following few weeks. The aircraft types included Army Skeeters and Austers, RAF Belvederes and Hastings and RAF Wessex. Many had good reason to be grateful for the assistance rendered to the public during 'The Big Snow'.

Sioux XT248 over Belfast City Hall at Christmas in the 1960s. *(RUC GC Foundation)*

Sioux XT812 being examined closely at RUC HQ Knock in 1964. *(RUC GC Foundation)*

The years from 1963 to 1969 were relatively quiet. The army aviation presence was maintained from November 1964 by the Queen's Dragoon Guards Air Squadron, which replaced its Skeeters and Austers with Sioux AH1 helicopters in August 1966. Austers served in the Province for nine years, from 1957 to 1966, Skeeters for six years, from 1960 to 1966. By the spring of 1969, six Sioux were being operated by the 17th/21st Lancers Air Squadron at Aldergrove (supported by 11 Royal Electrical and Mechanical Engineers (REME) technicians), with a further four helicopters from the Prince of Wales's Regiment being based at RAF Ballykelly. In September 1966, RN helicopter activity ceased at Eglinton.

One of the Phantoms delivered to No 23 MU from the USA on arrival at Aldergrove. *(Ulster Aviation Society)*

Perhaps the most dramatic arrival at RAF Aldergrove in 1968 was on 20 July when XT891, the first McDonnell Douglas F-4M Phantom fighter landed to receive the attention of No 23 MU. At this time, the Phantom was one of the most potent military aircraft in the world and was the newest type on the RAF and RN inventories. After acceptance checks and modifications to ensure in-service compatibility, the same aircraft was the first to be delivered to No 228 Operational Conversion Unit at RAF Coningsby on 23 August. Between 1968 and 1970, some 135 new Phantoms were handled by the MU.

Six Westland Wessex HC2s of No 72 Squadron fly over Aldergrove in August 1969. *(Ulster Aviation Society)*

On 14 July 1969 three aircraft and crew from No 72 Squadron, under the command of Squadron Leader Oliver, flew from exercise duties on Salisbury Plain to RAF Ballykelly in Co Londonderry, which at that time was still an important base for the Avro Shackleton fleet. On 17 August the CO, Wing Commander P Wilson, led a further three aircraft across and at that juncture all seven helicopters re-located to RAF Aldergrove, which was closer to Belfast. As the Support Helicopter Detachment forming the 'Ulster Flight' its main roles were security, trooping, re-supply and the movement of VIPs. Over the next two years the mainland commitments of the Squadron gave precedence to an escalating requirement in Ulster as the violence intensified, up to 12 aircraft at a time being based at Aldergrove.

?ersonnel of Scout Flight, 664 ;quadron pose with one of the ;couts at Aldergrove in 1971. *James Arbuthnott)*

Army vehicles at Sydenham 1971. *(Ulster Aviation Society)*

Scout XV132 lands at RUC HQ Knock. *(John Barnett*

Soon the increasing violence and civil disturbance resulted in the aviation assets being supplemented by the four Westland Scout AH1s of 8 Flight, which were sent to RAF Ballykelly in support of the troops and police in Londonderry, and also to Omagh in August 1969. The deployment of AAC assets expanded considerably as the mayhem grew even worse. 665 Squadron was deployed in Northern Ireland for the first time for a four month tour based at Aldergrove, beginning in June 1971. In the summer of 1972, six Sioux of the 16th/5th Queen's Royal Lancers were based at Omagh. Squadron, five more Scouts and three Sioux from 664 Squadron were at Aldergrove, Three of 664's Sioux were at Ballykelly, along with three more from 40 Commando Royal Marines. By 1973 some 15 Scouts, 23 Sioux and three Beavers were based in Northern Ireland. By 1977, five different locations were involved, Aldergrove on the eastern shores of Lough Neagh (1957 – the present day) Ballykelly on the north coast (1969–1991), Omagh to the west (1969–1982), Long Kesh south of Belfast (1973–1979) and, for a few years, Sydenham (1973–1977).

Buccaneers undergoing maintenance at RNAY Sydenham. *(Ulster Aviation Society)*

RAF Sydenham crest. *(Ulster Aviation Society)*

RAF Sydenham. *(Ulster Aviation Society)*

Hartland Point and Maidstone at Sydenham 1973. *(Belfast Telegraph)*

In 1969, Airport Wharf came into use again, this time as the berth for the prison ship HMS *Maidstone*. Sydenham became a tri-service establishment with a large influx of army personnel. On the other side of the airfield, a safe haven was provided from time to time for Belfast Corporation buses and bin lorries, which otherwise were a soft target for rioters. Along with the rest of the Province, the workers at Sydenham coped with the onset of renewed civil strife, the bombs and hoaxes, the shootings and strikes, while maintaining as much of the pattern of everyday life as possible. In July 1973, the RN relinquished the site to the RAF and so the station became Royal Air Force Sydenham and which in turn closed in 1978.

The RAF Ensign lowered for the last time at Ballykelly in 1971. *(David Hill Collection)*

Shackleton MR2 WR967 and Wessex HAS3 XP118 at Ballykelly in 1970. *(David Hill Collection)*

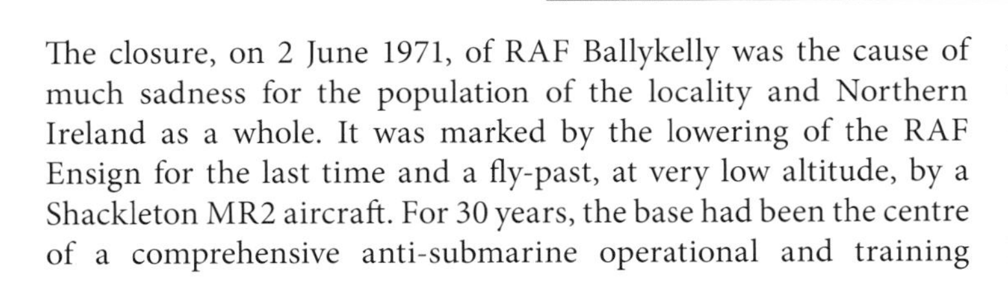

The closure, on 2 June 1971, of RAF Ballykelly was the cause of much sadness for the population of the locality and Northern Ireland as a whole. It was marked by the lowering of the RAF Ensign for the last time and a fly-past, at very low altitude, by a Shackleton MR2 aircraft. For 30 years, the base had been the centre of a comprehensive anti-submarine operational and training presence by the RAF and RN, dating back to the arrival of the Coastal Command Development Unit in December 1941. The airfield was handed over to the army who renamed it Shackleton Barracks, a tribute to one of the aircraft types, arguably the most significant, associated with it. The departure of the RAF did not, however, mean the end of flying from Ballykelly.

No 230 Squadron Pumas arrive at Aldergrove escorted by a Wessex from No 72 Squadron. Their flight path is taking them overhead Air Traffic Control and the Fire Station. *(Ulster Aviation Society)*

The arrival of four Westland Aérospatiale Pumas from No 33 Squadron at Aldergrove in December 1972 eased the burden on No 72 Squadron somewhat. April 1973 saw the start of a commitment to Northern Ireland by a detachment of Pumas from No 230 Squadron, which was shared with No 33 Squadron until 1979. October 1987 brought the start of further detachments to Northern Ireland by No 230 Squadron, which lasted until March 1990. This was, however, by no means the end of No 230 Squadron involvement with the province

Royal Marines Gazelle XX392 of Dieppe Flight, 3 Commando Brigade Air Squadron (3CBAS) in the grounds of Altnagelvin Hospital in April 1980. *(David Hill)*

The old and the new, Scout and Gazelle. *(Museum of Army Flying)*

Throughout the first half of the 1970s, until the arrival of the Westland Gazelle AH1 in 1976, Ballykelly was the base for a detachment of between three and five Sioux. The last Sioux to deploy to Ballykelly were from 665 Squadron, which departed in April 1977, sandwiched between the first two Gazelle detachments, provided by 651 and 659 squadrons. 655 Squadron became based at Ballykelly in 1982 and remained there as the resident AAC unit until 1991. Two Scouts of 8 Flight were sent to Omagh in August 1969. Thereafter, Scout detachments supported the Sioux of the resident armoured regiment (for example, 17th/21st Lancers, 16th/5th Queen's Royal Lancers, 1 Royal Tank Regiment, 13th/18th Royal Hussars, 15th/19th King's Royal Hussars, 9th/12th Royal Lancers) until 3 Flight, with firstly, Sioux and then Gazelles, took on the light helicopter role from 1976. With the formation of the Northern Ireland Regiment in 1979, 655 Squadron based Scout and Gazelles at Omagh until moving to Ballykelly in 1982, so ending 13 years of operational history.

Major Jeff Pink, OC 651 Squadron golfing at Long Kesh in 1975. *(via Mrs Olwen Pink)*

As we have seen, Long Kesh was a Second World War airfield, which later achieved notoriety under another name – HMP Maze. It functioned as a helicopter base for six years between 1973 and 1979. A mixture of Scout and Sioux helicopters gave way from 1976 onwards to a regular rotation of six Scouts and six Gazelles. The first tour of duty in Northern Ireland by 655 Squadron was carried out by Scout and Sioux helicopters, which were based at Long Kesh from February to June 1973, the first of four spells there, the last of which ended in June 1979. The premier AAC squadron, No 651, had detachments at Long Kesh from August 1974 to May 1975 and February to June 1978. Wherever in the world they have served, military personnel have always made time for sporting activities, as demonstrated here by Major Jeff Pink. The Squadron also experimented with keeping a small herd of goats. All went well until the goats raided the garden of the Officers' Mess.

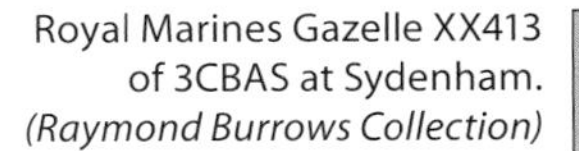

Royal Marines Gazelle XX413 of 3CBAS at Sydenham. *(Raymond Burrows Collection)*

Sioux XT202 operated with Kangaw Flight at Sydenham in 1973. *(Eric Myall)*

Sydenham (now George Best Belfast City Airport) was home to a succession of Royal Marine Commando and AAC Sioux detachments from February 1973 until August 1976, thereafter the final two tours were undertaken by the Gazelles of 3 Commando Brigade Air Squadron, until the withdrawal from this base in October 1977.

A wintery scene at Aldergrove in March 1980 with a Gazelle in the foreground and a Scout nearest the hangar. *(via Mrs Gabrielle Tait)*

REME at work at Aldergrove in 1980. *(via Mrs Gabrielle Tait)*

The deployment of helicopters and a few fixed-wing types for four-month tours of duty came under the title of Operation *Banner*. For a period of ten years between 1969 and 1979, when the Northern Ireland Regiment AAC was formed, the following AAC squadrons undertook tours of duty – some of them many times – 651, 652, 653, 654, 655, 657, 658, 659, 660, 661, 662, 663, 664, 665, 666 and 669. The major permanent location for the AAC in Northern Ireland has undoubtedly been Aldergrove with the succession of Austers, Skeeters, Sioux, Scouts, Beavers, Gazelles, Lynx, Islanders and Defenders from 1957 to the present day. As the earlier shorter range helicopters were replaced by more modern and capable aircraft it made sense to concentrate the resources at one location, which gave the advantage of centralised operational control, spares and servicing.

Four DHC Beavers from Aldergrove pictured on the cover of the 1986 *Army Air Corps Journal*. *(Museum of Army Flying)*

Northern Ireland last operational Beaver Flight in 1989. *(Ulster Aviation Society)*

The Beavers came on detachment firstly with a single aircraft only, early in 1972. The initial detachment of three aircraft was provided by 655 Squadron in October 1973. Further detachments were made by 3 Flight and 669 Squadron until the establishment of the Beaver Flight of five aircraft at Aldergrove in March 1976. The title 1 Flight had been dormant for 20 years until 1 October 1988, when Beaver Flight at Aldergrove was re-designated 1 Flight AAC.

A sad farewell to the Beavers was said some eight months later, the final flypast being made by XP769 and XP825 on 18 June 1989. The Beaver served in the Province for 17 years from 1972. They were employed chiefly on photographic reconnaissance and light liaison duties. The DHC2 Beaver was a robust and reliable aeroplane, which was well-liked by all those who flew in this great workhorse.

Sioux could even be armed with a machine gun – seen here over Magilligan. *(Belfast Telegraph)*

The last Sioux over Black Head lighthouse, Islandmagee on their departure from Northern Ireland in December 1977. *(Museum of Army Flying)*

The final deployment of Sioux consisted of those belonging to 652 Squadron, XT102, XT558 and XT815, which were based at Aldergrove from August to December 1977. Therefore the Sioux served in the Province for 11 years 1966 to 1977. The 'Clockwork Mouse' was a fairly basic machine, with a groundspeed not much in excess of 70 mph, comparable to a contemporary family car, which had the effect of making Northern Ireland seem like quite a large country! It could carry only a small payload and was not equipped with any sophisticated surveillance devices. The bubble canopy gave a marvelous all-round view but did not provide a happy feeling of armoured security from anything fired from the ground in its general direction. However, it performed a multiplicity of tasks including resupply of mail, laundry and water, the movement of personnel, the carriage at times of a police dog and its handler, surveillance by hand-held binoculars and nightly patrols of electricity pylons with a Nitesun searchlight.

A Wessex HU5 of 845 NAS over Londonderry in August 1980, undertaking an aid to the civil community task. *(Belfast Telegraph)*

A Wessex HU5 of 707 NAS RFA Resource Flight in 1977. *(Pat Hindley)*

1977 saw the deployment of naval Westland Wessex HU5s, firstly a trial by the RFA *Resource* Flight, which was part of 707 NAS. The aircraft operated from Sydenham. The first FAA unit to be based at Aldergrove since 774 NAS in 1939–40, was a detachment of 845 NAS Wessex HU5s in November 1977. These aircraft, which were very similar to the RAF's HC2s, spent much of their time either flying between Aldergrove and Bessbrook in Co Armagh or carrying out tasks from the base at Bessbrook Mill. Personnel served tours of six weeks in duration, in a two year tour this could mean five or six spells in the Province.

An English Electric Canberra and a Vickers Varsity at No 23 MU. *(Ernie Cromie Collection)*

Three Canberras and an Armstrong Whitworth Argosy at No 23 MU. *(Ernie Cromie Collection)*

No 23 MU, which had been at Aldergrove since November 1939, closed on 23 April 1978. By Northern Ireland standards, it had been a large civilian employer throughout its existence, though it is somewhat ironic that the employment total reached its all-time high of 1500–1600 only a matter of 10–15 years prior to disbandment. Persons departing from or arriving at Belfast International Airport are still able to see a substantial remnant of No 23 MU – the two large blue hangars near the fire dump just beyond where runways 07/25 and 17/35 cross.

Bessbrook in 1980. *(Ulster Aviation Society)*

Coming in to land at Bessbrook in a Gazelle. *(Guy Warner)*

The old mill at Bessbrook, which was first established as long ago as 1761, lies eight miles from the closest part of the border with the Irish Republic, 40 miles south-west of Belfast and 64 miles north of Dublin. It was the focal point of the Security Forces' operations in South Armagh and along the border with the Irish Republic since 1972. At its peak in the 1980s RAF, AAC and RN helicopters amassed over 600 flights a week in and out of Bessbrook – on average one every eight minutes during daylight hours and as many as 15,000 passengers a month, making it the busiest heliport in the world. The surrounding countryside has been described by a visiting pilot as, "a haphazard patchwork of small fields, laced by thorny hedgerows and tumbling streams, dotted with lakes and muddy pools and bounded by craggy uplands." It is overlooked by the Craigmore Viaduct which has carried the Belfast to Dublin railway line since 1852. Rising to a height of 126 feet, it has been an excellent landmark for many a helicopter pilot seeking Bessbrook in bad weather.

655 Squadron's Scouts arrive at Aldergrove to join their Gazelle colleagues and 654 Squadron Lynx AH1s to form the Northern Ireland Regiment AAC in October 1979. *(Ulster Aviation Society)*

A Lynx AH1 of 651 Squadron arrives at the scene of a bomb alert in 1980 on the M1 Motorway near Dungannon. *(via Mrs Gabrielle Tait)*

The Westland Lynx AH 1 was introduced in October 1979 with the deployment of aircraft from 654 Squadron to Aldergrove on a four month tour of duty. The next major change was the formation of the Northern Ireland Regiment AAC on 1 November 1979. Instead of 16 AAC squadrons rotating on tours of duty through the Province (some as many as seven times in ten years) proper provision was made to enable greater efficiency in the use of the resources available. BAOR AAC Regimental HQ involvement also ceased, with the withdrawal of 4 Regiment from Aldergrove in December and the last detachment of Scouts on roulement from Germany. The rotation of squadrons did not cease but the pressure was eased by the permanent presence of 655 Squadron. The first aircraft to join the new regiment were the four Scouts of 655 Squadron's Scout Flight, which arrived from AAC Topcliffe at the end of October. In the same year another significant event was the arrival of the rest of the Scouts and six Gazelles of 655 Squadron, with these helicopters being based at Lisanelly Barracks, Omagh. During the first three years of its existence the Northern Ireland Regiment consisted of the Beaver Flight, the 655 Squadron Scout Flight and the roulement squadron of Lynx and Gazelles – all based at Aldergrove, together with the Scouts and Gazelles of 655 Squadron at Omagh. Three Gazelles of the Royal Marines at Ballykelly also came under the control of the regiment. The aircraft were maintained and supported by 70 Aircraft Workshop Detachment REME and 3 Mobile Stores Detachment RAOC, both located at Aldergrove. The Scouts were replaced by Lynx AH1 helicopters and 655 Squadron moved to Shackleton Barracks at Ballykelly, Co Londonderry in the summer of 1982. The next type to depart from operational duties was the Westland Scout in October 1982. The last Scout task in Northern Ireland was carried out by Major MA Jones in XT623, flying guests back from Aldergrove to HQ at Lisburn. The Scout served for 13 years in Ulster from 1969.

No 72 Squadron aircrew relax in the Mess at Aldergrove. Note the framed photograph of the Squadron standard between two of the pilots. *(Guy Warner Collection)*

On November 12, 1981, Wing Commander AE Ryle AFC led a formation of the final three No 72 Squadron Wessex from Benson and so brought the Squadron's Standard to Aldergrove. Later that day, Wing Commander AA Nicholson MVO formally assumed command of the unit and its full compliment of 13 aircraft and crews. The Falklands war in the spring of 1982 caused a brief, partial re-deployment of the Squadron back to RAF Benson but it would soon increase in size at Aldergrove to 21 aircraft, 39 pilots and 123 ground crew, making it the largest operational squadron in the RAF.

This view gives some idea of the conditions faced by the rescue services during the *Antrim Princess* emergency and (inset) evidence of a fortunate escape for S/L Waldron and his crew. *(Guy Warner Collection)*

The most notable rescues carried out by No 72 Squadron were perhaps those of 9 December 1983. Firstly, the Royal Navy patrol vessel, HMS *Vigilant,* broke down half a mile from the Co Down coast near Donaghadee. The ship had to be taken in tow by local fishing boats, while the 26 crew members were winched to safety. This was merely the beginning, as with storm force 10 winds lashing the shore and a sea state of 8, the Larne–Stranraer car ferry, the *Antrim Princess*, suffered an engine room fire which caused it to drift powerless and in great danger a mile off Larne. This time no less than 108 passengers and 20 crew were lifted to safety. Four Wessex and five naval Sea King helicopters (from Prestwick) were involved, as also was a Nimrod flying overhead to co-ordinate operations. The dangers inherent in life-saving missions were dramatically illustrated when a rocket line fired from another ferry wrapped itself around one of the rescue helicopter's main and tail rotors. Very fortunately, the aircraft made an emergency landing on the nearest land. Several members of the squadron received gallantry awards as a result of their actions that day. Flying Officer Duncan Welham was involved in both rescues, flying as co-pilot with Squadron Leader Carlyle, firstly in XV728 and then in XR517 (now part of the Ulster Aviation Society's collection at Maze/ Long Kesh). He remembers vividly his first sight of the storm-tossed ferry, with other vessels clustering round – one of which fired the line which could have proved fatal to Squadron Leader Al Waldron and his crew. The whole tail resembled an orange cobweb. After spending over six hours in the air, several of which were spent helping the crewman haul survivors into the cabin, Duncan enjoyed his beer that evening – some months later he was surprised and pleased to receive £50 salvage money.

Sub-Lieutenant (A) Peter Lock in his dress uniform. *(Peter Lock)*

The UAS Wildcat being recovered from Portmore Lough by a Lynx of 655 Squadron AAC. *(Ernie Cromie Collection)*

Many aspects of the history of military aviation in Ulster are featured in various ways within the Ulster Aviation Collection at Maze/Long Kesh. One in particular is the restoration of Grumman Wildcat JV482, which has been referred to earlier. The engine and airframe, somewhat ravaged by natural forces and the scavenging antics of souvenir hunters during the 40 years that it lay in Portmore Lough, were raised out of the mud during the autumn and winter of 1983/84 by members of the Ulster Aviation Society,. They had marvellous assistance from numerous friends and organisations including the Heyn Group, Belfast and Ulster Sub-Aqua Club but the remainder of the Wildcat was eventually lifted off a temporary platform in spectacular fashion by a Lynx helicopter from 655 Squadron Army Air Corps, then based at Ballykelly. Not long afterwards, the pilot who had successfully ditched the aircraft after its engine went on fire, Peter Lock, was tracked down to his home in Canada, to which country he had emigrated from England shortly after the war. With his wife Marjorie, he returned to Northern Ireland to help celebrate the recovery of the aircraft and subsequently returned to Long Kesh where he and the aircraft had been temporarily based in December 1944.

665 Squadron Gazelles at Palace Barracks. *(Guy Warner)*

Gazelle over Carrickfergus. *(Guy Warner Collection)*

665 Squadron Gazelle over Belfast. *(Guy Warner Collection)*

In 1986, 665 Squadron was reformed at Aldergrove and became part of the Northern Ireland Regiment, equipped initially with a mix of Gazelles and Lynx. A new badge was conferred to it in November 1987. On it was depicted a single maple leaf in honour of the Squadron's Canadian origins, superimposed with a harp and the motto Providence to signify its deployment in Northern Ireland. It became a Gazelle only squadron in 1991. With 21 aircraft it was the largest in the AAC, it was also the busiest, flying some 12,000 hours per year. The Gazelle holds the record for length of service, having been in Northern Ireland for more than 35 years.

A 655 Squadron Lynx at Aldergrove with 665 Squadron Gazelle in the background. *(Guy Warner Collection)*

A 655 Squadron Lynx with its door gunner in position. *(Guy Warner Collection)*

Guy Warner preparing for a flight in a 655 Lynx, note the NVG on his flying helmet.

In 1991 655 Squadron moved from Ballykelly, joining 665 Squadron at Aldergrove. It standardised on the Lynx. The Lynx AH7 had been introduced in 1991, its main advantages over the AH1 were improved avionics, reduced noise, better hover capability and more advanced composite rotor blades. The Lynx in its AH1 and AH7 models served in the Province for almost 30 years.

A Chinook brings back a stranded 655 Squadron Lynx from Portrush, where it had suffered mechanical failure, necessitating a landing on the beach. *(Guy Warner Collection)*

On 7 April 1993 Chinook ZD574 assisted the Ulster Aviation Society in moving its Short 330 aircraft fuselage. *(Raymond Burrows Collection)*

655 Squadron Lynx, XZ667, departs from Bessbrook as No 7 Squadron Chinook, ZA704, prepares to lift an underslung load. *(Museum of Army Flying)*

The long-standing association between the Ulster Aviation Society and the military flying services in Northern Ireland was further demonstrated in spectacular fashion on 7 April 1993 by a RAF Chinook helicopter which airlifted the second Shorts SD3-30 pre-production prototype airliner from Nutts Corner to join the Society's expanding aviation collection which was then based as Langford Lodge airfield. The mighty Boeing Chinook or 'Wokka, Wokka' had an unparalleled lifting capability. The first to visit the Province was the HC1, ZA681/FZ, of 240 OCU for a fortnight in May 1982. Regular detachments of Chinook HC1 and, later, HC2 from No 7 Squadron were made from the late 1980s, throughout the 1990s and beyond. Tragically on 2 June 1994, in the worst military air crash of the troubles, Chinook HC2 ZD576 was lost on a flight from RAF Aldergrove to Scotland, with loss of 29 lives.

Aid to the civil community, a Puma and crew from No 230 Squadron assist the pupils of Belfast Royal Academy in cleaning up litter in the Mournes. *(John Reilly)*

As a proud member of the NATO Tigers Association, Pumas have often appeared resplendent in temporary colour schemes. *(via Wing Commander 'Harry' Palmer)*

Puma ZA939 'Tiger Scheme'. *(Raymond Burrows Collection)*

In May 1992, No 230 Squadron and its Pumas relocated to RAF Aldergrove, where in October a ceremony was held to present a new Standard to the Squadron. This was followed in August 1993 by a celebration of the Squadron's 75th Anniversary. Following its arrival in Northern Ireland seven Pumas of No 230 Squadron took part in Operation *Christo* which also involved two Chinooks and six Wessex, supporting the RUC and the Army in Crossmaglen, South Armagh, where they were investigating a smuggling racket.

A Puma with underslung load at Bessbrook.
(Guy Warner Collection)

A Gazelle landing at Bessbrook.
(Guy Warner Collection)

On 26/27 November 1992 as Gazelle ZB681 of 665 Squadron AAC was taking off at Bessbrook, it collided with an incoming No 230 Squadron Puma, XW233. The Puma impacted into the perimeter security fence, while the Gazelle crash-landed close by. The two Gazelle crewmen escaped with serious injuries but all four personnel on board the Puma were killed Squadron Leader M Haverson, Flight Lieutenant SMJ Roberts, Flight Sergeant JR Pewtress and an Army officer, Major J Barr, who was on a familiarisation sortie. This was the first tragedy of this nature at Bessbrook in over 20 years of operations.

A Gazelle of 5 Regiment proves a great attraction for local mini-rugby teams on a visit to the Belfast Royal Academy playing fields at Roughfort. *(Guy Warner)*

The Gazelle departs from Roughfort. *(Guy Warner)*

On 1 October 1993, the Northern Ireland Regiment AAC was retitled 5 Regiment AAC. As well as many important military tasks, over the years its helicopters have also carried out an important PR function in the community at large. Indeed the Commanding Officer of 5 Regiment, Lieutenant Colonel Chris Butler, joined the Academy mini-rugby club in order to obtain his coaching qualifications. The Gazelle is well-named, it is graceful, agile and a delight to fly, though proving a challenge to operate at night on single-pilot missions.

A pair of Sea King HC4s in South Armagh. *(via David Green)*

A Commando Sea King HC4 with an underslung load in South Armagh. *(via David Green)*

From 1 October 1993 to 31 March 1994 Sea King HC4s of 707 NAS were based at Aldergrove. Then on 1 April 1994 Sea King HC4s of 846 NAS arrived at Aldergrove. They performed the same function as the RAF's Wessex and Pumas in support of security forces personnel on the ground. The naval deployment ceased in March 1999, having been at Aldergrove for five years. A story is told that on one sortie in South Armagh a policeman and his dog were being flown to an incident. The friendly aircrewman offered the dog an extra-strong mint, which he gobbled up. That, as the dog's handler pointed out, was the end of its usefulness for the day, as his sniffer dog could now only smell mint!

The Buccaneer arrives at Langford Lodge with Gartree church in the background. *(Ulster Aviation Society*

994 BAe Buccaneer XV361, which had been purchased
er Aviation Society, flew from RAF Lossiemouth, firstly
dergrove and then, on 5 April the short hop to join the
ation Collection which was then located at Langford
at 1 minute 31 seconds, the shortest ever flight by this
aeroplane. Royal Navy Buccaneers had previously be
visitors to RNAY Sydenham for maintenance until 19
served with the RN for 14 years before transferring to th

No 72 Squadron Wessex and the Spitfire *Enniskillen*. *(Ulster Aviation Society)*

No 72 Squadron 25th Anniversary at Aldergrove in 1994. *(Ulster Aviation Society)*

In September 1994, No 72 Squadron celebrated 25 years of continuous service in the Province, flying the Westland Wessex. The commemorations began with the Squadron Standard being paraded before the Secretary of State for Northern Ireland, Sir Patrick Mayhew, while three Wessex flew past with the RAF Ensign in tow. Also present on the day were the Battle of Britain Memorial Flight Spitfire IIA P7350, painted to represent the No 72 Squadron aircraft *Enniskillen,* the Red Arrows and the Army Helicopter Display Team, the Blue Eagles. During the Second World War the *Belfast Telegraph*'s Spitfire Fund raised sufficient money by public subscription to purchase 17 Spitfires for the RAF.

JHFNI Lynx and Gazelle. *(Museum of Army Flying)*

A Joint Helicopter Force Northern Ireland (JHFNI) trio – Wessex, Puma and Chinook. *(Ulster Aviation Society)*

15 August 1998 was the date of the dreadful Omagh bombing, helicopters from 72 and 230 Squadrons, as well as a Sea King HU Mk 4 of 846 NAS, ferried casualties to hospitals in Enniskillen, Londonderry and Belfast. Twenty-nine people were killed and 399 injured. The Joint Helicopter Force Northern Ireland was established at Aldergrove on 1 April 2000 (though in effect it had been functioning since the previous October). This brought all service helicopters in the Province under one command and control authority.

A No 72 Squadron Wessex exercises with the RNLI. *(Ulster Aviation Society)*

Wessex formation lifts and departs from Aldergrove for the last time. *(Paul Harvey)*

A No 72 Squadron Wessex over Carrick-a-Rede rope bridge. *(Ulster Aviation Society)*

In March 2002 No 72 Squadron stood down after 33 years of service in Northern Ireland. No 230 Squadron became the sole RAF Support Helicopter unit in Northern Ireland. Wessex HC2, XR517, was purchased by the Ulster Aviation Society in 2003. It served in Northern Ireland with No 72 Squadron for 20 years and is now part of the Society's Heritage Collection at Maze/Long Kesh.

A Lynx of 815 NAS at Bessbrook. *(Guy Warner)*

Two 655 Squadron Lynx. *(Museum of Army Flying)*

In March 2007 655 Squadron stood down and some of its Lynx were transferred temporarily to 665 Squadron. Part of the capability gap lost by this and also the withdrawal of 230 Squadron RAF's Pumas from tasking in Northern Ireland (so its aircraft and crews could be re-deployed to Iraq) was taken up from 6 September to 7 December 2006 and then again from 20 March to the end of July 2007, by up to three Lynx HAS3 of the 815 NAS Operational Readiness Unit (ORU), which were based at Aldergrove in support of the Police Service of Northern Ireland (PSNI). The role of these Lynx was to act as the daytime province utility aircraft. This consisted chiefly of the insertion and collection of patrols in South Armagh and also the secure transportation (by underslung load) of blasting explosives for quarry operators.

The last Army helicopter to land at Bessbrook. *(Guy Warner)*

Chinook ZH891 at Bessbrook. *(Guy Warner)*

The 31 July 2007 saw the end of 38 years of Operation *Banner* in Northern Ireland. One of the authors of this book, Guy Warner, had the great privilege of spending a day with a Chinook and its crew, while it flew sorties from Bessbrook assisting in the dismantling of the hilltop observations posts and the restoration of the landscape to its original condition. He also flew as a passenger in the last Army aircraft to land and take off from Bessbrook Mill – Gazelle AH1 ZB689 on 21 June between 10.15 and 10.55, with Captain Ian Cameron and Staff Sergeant Jase Wright. For the historical record, the last aircraft to leave from Bessbrook was the Royal Navy Lynx HAS3, XZ 735, crewed by Lieutenant Marty Craven, Lieutenant Rae McDermott Evans and Lieutenant Mat Askham, which lifted from the HLS at 10.05 local on 22 June 2007, arriving back at Aldergrove at 10.25 local, so bringing to a close an era that will have many memories for aircrew of several generations.

Defender Mk 2 ZH001 undergoing maintenance at Aldergrove in 2009. *(Guy Warner)*

Islander AL1 of 651 Squadron with a Gazelle of 665 Squadron in the background. *(Museum of Army Flying)*

The first army Britten-Norman Islander AL1 ZG846 had arrived at Aldergrove on 10 March 1989, a flight of five aircraft being formed, taking on similar duties to the Beaver, which it replaced. 1 Flight was subsumed by 651 Squadron, which arrived at Aldergrove in August 2008 and also operated the more capable development of the Islander, the Britten-Norman Defender 4S Mk 1, 2 and 3. The Islanders and Defenders have both served overseas: the Islander in Saudi Arabia in 1991 during the First Gulf War and the Balkans later in the decade, while the Defender has operated in Iraq (Operation *Telic*) and Afghanistan (Operation *Herrick*).

A 665 Squadron Gazelle overflies Dunluce Castle. *(Ulster Aviation Society)*

Islander AL1 ZG848 of 651 Squadron. *(Ulster Aviation Society)*

Defender Mk 3 ZH004 flies past the Mussenden Temple. *(Ulster Aviation Society)*

On 20 September 2009 the RAF Ensign was lowered at RAF Aldergrove. It now became Joint Helicopter Command Flying Station Aldergrove with the sole military aviation presence being 5 Regiment AAC, with its Islanders, Defenders and Gazelles in the Manned Airborne Surveillance Role. The first Army Air Corps officer to become Station Commander, Colonel Richard Leakey, arrived in post on 1 September 2010.

Three of No 230 Squadron's Pumas lift off from Aldergrove for the final time. *(Paul Harvey)*

A fine study of a Puma in flight over Ulster's lakeland. *(Ulster Aviation Society)*

No 230 Squadron leaving Aldergrove on 17 October 2009. *(Ulster Aviation Society)*

Autumn 2009 brought the departure of No 230 Squadron which had been based in the Province since 1992.

A QUAS Chipmunk in the early 1960s. *(via Liz Shanks)*

Grob 109B Vigilant T1, ZH 209, of the Air Training Corps lands at Langford Lodge. *(Eric Gray)*

Landing at Bishops Court. Cadet Andrew Irvine with motor glider Slingsby Venture T2, ZA629. *(Andrew Irvine)*

The Air Training Corps (ATC) and the Queen's University Air Squadron (QUAS) have been an important means of introduction to aviation for many young people throughout Northern Ireland. The Squadron was formed on 8 January 1941 at Sydenham within No 54 Group and was redesignated Queen's University Air Squadron on 18 May and disbanded in January 1946. It reformed at Aldergrove in October 1946 under the control of RAF Northern Ireland. In March 1947 it moved to Sydenham before returning to Aldergrove in 1992, where it remained until disbanded on 31 July 1996. The aircraft used included DH Moth, Avro Tutor, DH Tiger Moth II, North American Harvard T Mk 2b, Airspeed Oxford, Hunting Provost T Mk 1, DHC Chipmunk T Mk 10, Avro Anson T Mk 21, Percival Prentice T Mk 1 and Scottish Aviation Bulldog T Mk 1. From December 1958 it also parented No 13 Air Experience Flight. The ATC today consists of 17 squadrons throughout Northern Ireland and for the last 70 years it has given both training and enjoyment to many thousands of young people. In 2012, 664 VGS (Volunteer Gliding School) remains at Newtownards with the Grob Vigilant T1 motor glider, serving the Northern Ireland Wing of the Air Training Corps and the RAF sections of the Combined Cadet Force. It was originally set up in August 1986 at RAF Bishops Court with one Slingsby Venture T2 motor glider, going into suspended animation in 1990, before re-forming at Sydenham in 1996, flying from there, Ards and Ballykelly until a permanent move to Ards in 2000.

QUAS Bulldogs at Sydenham. *(Guy Warner Collection)*

A damaged Bulldog is airlifted by a Puma in January 1997. *(Guy Warner Collection)*

QUAS Chipmunk WK567 sadly ended its days on the fire dump at Sydenham. *(Ernie Cromie Collection)*

Alouette 196 at Musgrave Park Hospital on 10 June 1970. *(John Barnett)*

Alouette 197 at Langford Lodge – the authors are first on the left and second on the right! *(Ulster Aviation Society)*

Friends and neighbours – The Irish Air Corps

Brigadier General Paul Fry, the GOC of the Irish Air Corps notes that military aircraft from the Republic of Ireland have been regular visitors to Northern Ireland over the years, firstly in the summer of 1949, when Avro Ansons from Baldonnel and Vickers Supermarine Seafires out of Gormanston, were tasked to carry out an extensive series of photographic flights. The Ansons undertook the vertical photography from a height of 5000 feet, while the Seafires provided the low-level oblique coverage from 800 feet. Assistance was given by the authorities in Northern Ireland with re-fuelling being carried out on occasion at the RNAS, Eglinton, where, to save time, Wrens brought sandwiches to the pilots in their cockpits. A letter from the Royal Navy Survey Section in London complemented the Air Corps on its photography, "twice as good as anything I have ever seen from the RAF." During the Troubles, there were regular amicable encounters when RAF and AAC aircraft undertook border patrols, as Paul Fry recalls, "There is a stretch of land to the west of Jonesborough where it would look as if both of us were on the wrong side of the border as the frontier runs north/south and our side is the east of the dry stone wall running along the summit of a spine of rock hills and your side lies to the west of it. On one occasion, I also encountered a formation of two Phantoms out in the Crossmaglen area trailing a large amount of smoke (which was what attracted my attention to them) whilst they were engaged in a very fast, very steep turn to avoid crossing over to our side!" Paul also undertook air ambulance missions in the 1970s and 1980s, flying Alouette IIIs, "The Royal Victoria Hospital in Belfast was my destination in 1979 for my first one, taking a seriously injured road crash victim there from Sligo Hospital. During refuelling at Aldergrove there was a small problem. The Army Air Corps bowser had run out of fuel just before I arrived and they had asked the RAF to send over theirs to top them up. We all had some difficulty when they tried to reconcile the fuel transferred to the Army bowser and then to me, as the RAF had issued the fuel to the Army in kilograms; the Army bowser had delivered it to the Alouette in pounds and my fuel gauge was calibrated in US gallons! We

Dauphin No 246 at Langford Lodge. *(Ulster Aviation Society)*

PC9s break. *(Guy Warner)*

Irish Air Corps AW139 taking part in an exercise in Northern Ireland. *(Irish Air Corps)*

PC9s display at Newcastle in 2011. *(Guy Warner)*

eventually sorted it out with a navigation computer. The Dauphins used Altnagelvin Hospital in Derry more frequently as it was closer to Finner and Sligo in the SAR days." The Alouette III, 202, was presented to the Ulster Aviation Society by the Irish Air Corps and is now on display at Maze/Long Kesh. Occasional visits have also been made to airshows; the Fouga Magisters of the Silver Swallows team flew at least one display at Newtownards and the Pilatus PC9 team took part in the display at Newcastle in 2011 and 2012. Dauphin helicopters took part in shows at City of Derry Airport, while several types, including the impressive CASA 235 maritime patrol aircraft, graced Ulster Aviation Society Open Days at its former home, Langford Lodge. Politicians and civil servants were brought to Belfast during the negotiations which resulted in the Good Friday Peace Agreement and, accordingly, Ministerial Air Transport Service (MATS) types such as the Beech King Air, Grumman Gulfstream IV and Bombardier Learjet 45 have been regular visitors. In recent times joint exercises with military and civilian services have been completed at Ballykinler and Magilligan Camps, practicing roles in emergency evacuation in a civil support scenario.

INDEX

People

Airfields, landing grounds and flying-boat moorings

Squadrons and Units

Royal Air Force:

Royal Navy and Royal Marines:

Army:

US:

Aircraft